German secret weapons: blueprint for Mars

D0533645

German secret weapons: blueprint for Mars

Brian Ford

Pan/Ballantine

Editor-in-Chief: Barrie Pitt
Art Director: Peter Dunbar

Military Consultant: Sir Basil Liddell Hart
Picture Editor: Robert Hunt

Executive Editor: David Mason
Designer: Sarah Kingham
Cover: Denis Piper
Research Assistant: Yvonne Marsh
Cartographer: Richard Natkiel
Special Drawings: John Batchelor

Contents

The war of lost opportunities

Introduction by Barrie Pitt

This fifth 'Weapons' book in our series is a notable contribution to a still largely untold story. German technology has always been a potent factor in world affairs and other nations have ignored the advances of German scientists, in peace and war, to their cost; the Allied propaganda campaign at the beginning of the war which tried to convince us all that no original creative thought ever sprang from German imaginations, and that all their scientific progress was but a pallid copy of our own, was not only singularly inept; it was basically unsound and unsupportable.

Not only was a considerable body of basic scientific invention conceived east of the Rhine, but the practical application of German scientific progress towards the solution of industrial problems had reached a stage much in advance of anyone else's. As a result, as Brian Ford convincingly demonstrates, the Nazi war machine swung into action utilising as much as it could of the most up-to-date scientific knowledge available, and as the war developed the list of further achievements grew to staggering proportions. From guns firing 'shells' of air to detailed discussions of flying saucers; from beams of sound that were fatal to a man at fifty yards, to guns that fired around corners and others that could 'see in the dark' – the list is awe-inspiring in its variety.

Fortunately for us, political factors prevented the German war machine from using much of its latent expertise as efficiently as it might have done. In some fields, events forced a 'stop-go' policy towards many of the branches of technological developments, with the result that projects which had been intensively pressed forward to within striking distance

of success, were suddenly relegated to obscurity as a result of some overall change of strategy or policy at government level.

But as readers of this book will learn, though the actual attainments of the German experts were uneven - some of the advances were very much less developed that we have imagined - by 1945, some were dangerously near to a completion stage which could have reversed the war's outcome. Not the least fascinating part of the book deals with some of the factors which saved the free world from domination by Hitler's megalomania, and there are surprising conclusions to be drawn; the parts played by various clauses in the Treaty of Versailles in combination with certain facets of the German temperament, are lucidly explained and well-argued - though doubtless some of Brian Ford's conclusions will create controversy in both scientific and historical circles.

But the outcome is clear; the Nazis were advanced in their attitude to science and scientific thought, and they were ready to develop as much as they felt necessary for victory in the way of scientific manpower and resources. They made some cardinal mistakes - and for these we should be eternally grateful, for they protected us from developments which would almost certainly have turned the balance of victory against the Allies. We would do well to remember that it was not the Blitz which began the proposals for complete evacuation of London, but the arrival of the V-bombs - the first of Germany's secret weapons to strike.

There were many developments beyond this, about which many people have never heard until now - but they could have cost us the war.

The centres of German research

The German character has always respected practical attainments and academic endeavour. To this day, the visiting industrialist who goes to Germany – West or East – finds how helpful it is if he admits on his visiting card that he is 'Mr Engineer' or 'Herr Doktor'; education, learning, and academic status have always been important parts of the German tradition.

In the 1930s this tendency was developed to the full. Through the propaganda machine of the Nazi empire-to-be, the academic and the engineer alike were esteemed as never before, and the aim of all successful men was to enter these professions and succeed within their framework. But as the Hitler regime came to power and began to exert its influence, there was a subtle indeed almost barely detectable change of emphasis. The pure scientist began to lose out of the favourable comment; the academic lost a little in favour – but the technician, the practical man, the engineer, these began an unprecedented climb to the greatest heights of status.

The shift of emphasis became outright bias, however, and particularly as more and more German scientists were being discriminated against because of supposed 'racial inferiority', many of them uprooted themselves and fled the country altogether. By the late 1930s the change had been almost complete: only Göring remained with any deep respect for the intellectuals of Germany, and he used them to the full. One of his chief co-workers was a General Milch, part-Jewish, who became Head of the Technical Office of the Luftwaffe in due course. In spite of 'mongrel' background, as defined by Hitler, Göring had this man kept in a senior position for pure intellectual ability and practical skill.

But to some extent this anti-intellectualism of the Hitler regime did have its desired beneficial effect, for it turned the German people away from their almost slavish acceptance of the need for academic specialisation, and allowed them to assume that (because of the widely-publicised 'inherent superiority' of the German race) they were above the need to specialise: they could all be convers-

Testing an A-3 rocket in 1937, an ancestor of the V-weapons

ant with the problems of technology and the scientific society, and great pains were taken to make them feel that – no matter how superficially – they were *in* on things. Secondly, because of the drift from academic endeavour, more and more people became technical workers, and the shift from pure research was accompanied to a certain extent by a drift towards applied research, design, and development. The cult of progress became established, and in the German mind it was readily nurtured.

Germany has an equal tradition for good quality workmanship, for discipline and for endeavour. Thus it was that many of their largest firms were in the export field, with singularly up-to-the-minute sales equipment to back them, and this – prophetically – included the development of munitions. The wheels of big business soon allowed this side of the German industrial endeavour to reach large proportions; the Germans were one of the few nations who were in a position to supply modern, effective munitions. Why was this? Quite simply because of their active research capacity: munition supply is one of the branches of industry which, almost more than anything else, relies on being up-to-date – in short, the successful munitions manufacturer must be the most advanced technically. This and the encouragement of militarism by the Nazis as an ideal led inevitably to the upsurge of successful, giant, weapon-manufacturing complexes.

And there was another factor, too, which – though designed to put a brake on the Germans' rearmament and to slow down their capacity to develop new weapons – actually had the effect of greatly intensifying development. This was the Treaty of Versailles which forbade the production of large ships, of high-capacity aircraft, of large-calibre weapons; but the Germans quickly overcame these limitations as far as they could by devoting new energies to making effective weapons *within* these limits. Thus one had convertible firearms, which could quickly be adapted for military use; one had high-velocity guns; one saw the pocket-battleship arise and the perfection of aircraft and gliders – all factors which, between them, enabled the Nazis quietly to evade many of the apparently inevitable restrictions of the Treaty of Versailles.

Factories in the industrial combines of Krupp, Mauser, and many others supplied arms and ammunition to many countries – including, in some cases, entire manufacturing establishments to countries as far away as South America – and including others, such as Russia, later to become her foes.

Even before the First World War there had been an Army Weapons Office, which had a branch known as 'Wa Prüf' – an abbreviation of *Heereswaffenamt Prüfwesen*, or Army Testing Office – designed specifically for the testing and improvement of weapons. It was, in essence, a proving ground and from it many important new changes and modifications were derived. One of the experts in this division, Carl Cranz, later formed a section of the Wa Prüf known as *Waffen Forschungs* – Wa F for short – which was specifically set up as a research and ballistics institute in its own right. This formed the first basis for further development in the Hitler regime; indeed when Cranz retired (aged over seventy, according to reports) he was replaced by a Professor Schumann and it was he who remained in charge right through to the end of the Second World War.

But here too the trend away from research for its own sake took a toll. for the institute became less prestigious and its leader found he was often left virtually out in the cold; it was the more practical activities of Wa Prüf which seemed to be in greatest demand. Thus it was that the munitions manufacturers who did not wish to incur the labour and expense of establishing their own research institutes, passed their work over to the Waffenamt – but found that the drift away from pure research tended to deny them many of the benefits they might otherwise have derived. So, in essence, the Ordnance did not have the research facility they needed. When eventually things did develop in this sphere, it was almost too late. However the practical experiences of fighters and tacticians using German weapons in the Spanish Civil War did provide some valuable practical trials and experience of the weapons in practice.

In the naval field, much important introduction of new technology was undertaken. The limits set by the Versailles treaty on warships was 10,000 tons; but by the maximum use of light-alloy materials and the development of high-rate arc welding of a remarkably sophisticated degree of design, the German technologists were able to overcome many of these limitations.

The research effort was largely based on the investment of considerable sums by the German business concerns who stood to make a killing by the production and sale of successful weaponry and equipment. There was an official *Marine-Waffenamt* (Naval testing office) under the Minister who acted as the Naval Commander – *Oberkommando der Marine* – and there were several experimental establishments *(Versuchsanstalt)* too. These included several organisations under the headings of *Chemische-Physikanalische* (Chemical and Physical Research), *Torpedo*, *Sperr* (Mines), and *Nachrichen* (Radio). Other facilities such as the *Forschungsentwicklung Patente* took care of patents and legal operations.

However in naval research too, in spite of the restrictions of Hitler's anti-intellectualism, the German resources were such as to establish a world lead in technical perfection and expertise. But in the Luftwaffe, things were somewhat different.

Here there was strong government research interest and, rather than leave things too much to the individual activities of the business combines, the technical competence of the government's resources was developed to a state of high activity and production. By shelving off the somewhat arbitrary demands of the policy co-ordinators of the government, the German air ministry was readily able to guard its independence of action; it would not be intimidated by anyone, and – probably partly as a result of the haughty, almost arrogant self-satisfaction of the army and navy research workers – it managed to create an aura of superiority for itself. Though Germany, for the reasons we have already outlined, had a justified reputation

Test firing one of the earliest rocket motors in July 1929

as a leading producer of artillery and naval equipment, there were many other countries with equal or better air ministries and Germany did not have any unique position of peerlessness in this field. But the high morale of the Luftwaffe paid off handsomely and indeed it enabled the Germans to achieve very advanced aims indeed. The rocketry research and development, as a case in point, was, as we shall see, remarkable and indeed quite unique as an exercise in the application of technology on an unprecedented scale.

It was in 1935 that Germany managed to escape from the strictures of the Treaty of Versailles and set about the redevelopment of her air force in a big way. Not that she came to the problem completely cold: a secret (and quite illegal) arrangement had been under way for some years before – exactly how many is by no means certain – by which German airmen had been instructed and aided by the Russian air force in a reciprocal agree-

ment. The Chief of Staff of the Luftwaffe at about this time, General Wever, was fanatical about the potentialities of larger and longer-range aircraft as part of the expansionist policy of the Nazis. It must have been with great satisfaction that Germany built and flew the first all-metal aircraft of any size at this time – the Dornier X – and many international trophies and prizes went to German aeroplanes in the late 1930s. It is said that a record speed of 469.22 mph was reached in April 1939 by a Captain Wendel, flying a Messerschmitt 109(R) – a speed not to be reached again until after the war's end, at least by airscrew-propelled aircraft.

Even in this field the Germans were working secretly on a number of projects which were later to surprise the Western world at large; jet-propulsion was at this stage very much more highly developed than the Allies knew, and rocket-powered aircraft were already on the drawing board. The most terrible of all of the German secret weapons were the rockets, of course – and these were beginning to be developed too, behind closed doors; as early

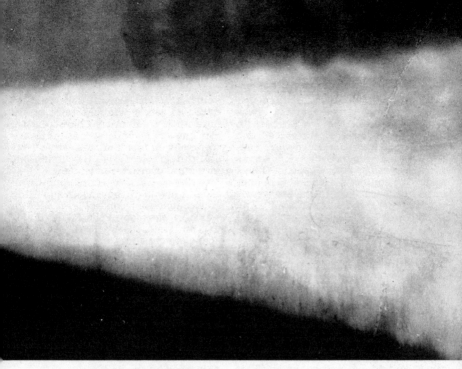

as 1931 the first of the modern liquid-fuelled rockets took to the air and reached a height of perhaps 1,000 feet from a base in Dessau and within two years secret teams were investigating the possibilities of manned rocket flight. The quickest way of reaching the enemy is through the air, and it is only natural that it was the Luftwaffe research establishments that were amongst the most progressive in forging these new, surprising weapons of war.

And so whilst the military and naval specialists worked for much of the war effort through the independent, business-backed organisations designed to develop new – and thence marketable – weapons, the Luftwaffe research remained close to the government. It would have been senseless to set up governmental establishments, when there were such clear risks of duplication of the independent laboratories, and in addition it would have been financially difficult to tempt away the industrial research workers – who were by this time amongst the most highly-paid technologists and designers in Europe, and probably in

the world.

But, with no traditional aircraft industry, the government became the only real supporter of aerial research; the men were trained, appointed, and distributed by a central machinery run by the Ministry at a senior level; their ultimate head, Göring, was as we have seen an admirer of brain-power and what it could attain; and as the years ticked by the developments themselves set a precedent which (though badly-organised and too spasmodic to be effective by modern standards) had not been seen before in the history of warfare. For its time it was incredible – and it worked.

But where were the establishments, and what were they like? Perhaps as important, just how was the organisation arranged for this mammoth task?

At the head of the army research was the Supreme Commander, who – through Speer's Ministry of Arms and War Production – controlled the general policies of the Wa Prüf. On a par with this department stood the *Waffen Forschungs*, weapons research section, which tended always to teeter on the brink of prominence but which (probably due to poor organisation and conflicting policy decisions as the war progressed) never came to hold the same degree of prominence as Wa Prüf. Many students of the war years have in fact imagined that Wa F was a sub-division of the Wa Prüf itself, but in organisational terms the two were of equal status. Both were controlled in a single office known as *Heereswaffenamt*, or Weapons Office, under the control of General K Becker until his death early in the war years, when General Leeb took over. And finally, working alongside the departments Wa Prüf and Wa F, was the *Beschaffung*, or purchasing and production section. This was the commercial division responsible for obtaining tenders for production, the buying of raw materials and the letting of production contracts to outside firms.

Subdivisions were set up to investigate such separate branches of research as ammunition and weapons, engineering – in the broadest sense – signalling, optical and communications equipment, and rocketry. This somewhat anomalous state of affairs arose because rockets were regarded (as they still are, by some military men) as having a split personality. Some say they are in essence artillery shells, which happen to take their cartridge charge with them; others argue that they are really aircraft but with shorter wings and without the pilot.

And so two divisions of the army's Wa Prüf were set up: one for solid-fuelled rockets, the other for liquid-fuelled. With an enthusiastic Major-General Dörnberger at the head, a team of some 250 of Germany's best young scientists was assembled before the outbreak of the war and they were given money, status and equipment to – simply – develop world-shattering rockets. From the pre-war site of Kummersdorf, the group moved in 1937 to *Heeresversuchsstelle* (army testing ground) Peenemünde and began work in earnest. Later the facility was dispersed to Bliecherode and Kochel, after the Allied forces had learned of the Peenemünde centre and begun to attack it.

Kummersdorf proving ground – situated near the capital Berlin – was then developed purely as a proving ground for rockets and guns. There were said to be fifteen separate test areas, but throughout the war period the facility was not stretched to capacity. Many of Germany's most up-to-date and secret weapons were tested here until their every characteristic was known and understood, and as the war went on much of this assessment and proving analysis was carried out at a similar ground at Gottow.

Chemical warfare, which might well have provoked the most appalling consequences of conflict ever seen in warfare, was also in the Nazis' minds at this time. As we shall see, they spent much time and effort in the pursuit of faster, deadlier poisons and developed, among other less sophisticated secret materials, several potent nerve gases by the war's end. The centre of development and testing was at a proving ground near Raubhammer. The whole enterprise was carefully controlled and the camouflaged buildings were often virtually undetectable to even the closest aerial reconnaissance by the Allies.

And backing the whole set-up were the educational establishments and colleges (the *Hochschulinstituten*) – over 200 of them – and the independent companies or *Firmen*, on whom much of the research depended.

The organisation in the navy was basically similar: here too there were separate sub-divisions of the parent Ministry office, and as in the army research, much of the effort relied on the cooperation and support of the independent companies. The relevant head office here was the *Marine-Waffenamt* (Naval Weapons Division) under Speer. The various specialised sub-divisions were similar to those of the army and they were in turn backed by the experimental and proof divisions. These provided a cybernetic feed-back link to the development divisions, since teething troubles and suggested improvements that came out of the proving tests were rapidly and efficiently absorbed into the rationale for the following phases of development and in this way – a form of mechanical evolution by 'survival of the fittest' – the quality was not only maintained but steadily and consistently improved.

The organisation of the air ministry was immense. In the very beginning of the preparation for war there was a change away from the organisational machine of the army and navy research in that Reichsmarschall Göring took a prominent personal stand at the top of the tree and had overall control of policy and development (even above the level of authority of the Ministerium Speer). Immediately below him there was a split into two functions: the Reich Luftfahrtministerium, or Air Ministry proper, and the scientific and technical branch, responsible for secret weapon development amongst other tasks.

One of the main divisions here was the Berlin-based *Technisches Amt*, the chief technical office of the Ministry itself. Initially at the head of this important division was General Udet; he was replaced by General Milch for the bulk of the wartime period and, later, by General Diesing. Most of the staff of this division were, in fact, military men and their task was basically to organise and co-ordinate research and development of aircraft, aerial weapons, communications equipment, and the like – all of it done under conditions of top security.

The separate specialised organisations themselves were varied. *Zelle* was the division concerned with airframe design; *Motor* handled the production and research into aeroplane engines of all kinds. *Geräte* (instrumentation) and *Funk* (radio-communications and radar equipment) supplied the most up-to-date equipment for the flying forces, and *Waffen*, or weapons, carried out a prodigious amount of development into armoury of all kinds, with the exception of bombs. This was the responsibility of the *Bomben* division, who also had the assignment of developing new bomb sights and aiming equipment. *Boden* handled ground-based equipment and *Torpedo* included the research into mines dropped from aircraft of all kinds. The *Fernsteuer Geräte* embraced the rocketry that led to the development of the V-1 flying bomb. This was simply because, as described earlier, some of the rockets were regarded as being 'pilotless aircraft' and, as such, clearly they ought to be placed under the Air Ministry rather than those

15

which (like the V-2) were essentially wingless missiles. This did mean, though, that there was a fundamental division between the two activities.

The whole operation was co-ordinated through the *Forschung führung* (literally meaning research-guidance) division, generally known as Fo-Fü. Its team of four scientific chiefs was always on hand for discussions with the Berlin powers and the degree of co-ordination effected between research and requirements was great – too great, as it turned out, for changes of emphasis at governmental level were often rapidly transmuted into a sudden alteration in a research programme which, whatever might be argued about its short-term expediency, cannot have done any good at all to the progress of the overall effort.

And finally, acting as the workhorse of the whole machine, there were several *Anstalt* establishments under the supervision of a director who controlled the several separate units in each institute. The Fo-Fü had laid down a policy on the establishment of such institutes, which laid stress on congenial fraternal control, good living standards, and a dignified working environment; plenty of finance and material backing and an opportunity for the frequent exchange of ideas on the interdisciplinary basis so necessary for effective furtherance of high-rate research.

The *Zentralstelle für wissenschaftliche Berichterstattung* (Centre for Scientific Records) acted as a centre for the co-ordination of publications of new discoveries. All scientists – even those working in secret fields – like to see their work in print, and numbers of reports were produced and circulated to personnel who were involved. A number of special yearbooks were instituted to bring the recognition of leading scientists to the attention of their more distant colleagues. Much was done to raise morale and efficiency – and it paid off handsomely in many respects. So, come to that, did the positions held by the scientists: salaries equivalent to $5,500 (£1,830) were paid annually to a typical research-worker, and that was worth vastly more in Germany at that time than it seems to be in today's terms.

Let us take a look back at the kind of surroundings that these scientists worked in – they were remarkable, even by today's standards, and have a distinctly James Bondian aura about them.

On the outskirts of Braunschweig lay a large area of woodland, surrounded, in the more open countryside, by a few scattered farm buildings. At least, that is how it appeared to aerial reconnaisance. But this innocuous little corner of Germany was actually something quite different – underneath the camouflage. This was the *Luftfahrtforschungsanstalt Hermann Göring*, the Göring Aerial Weapon Establishment, and it was one of the leading centres of top-secret developments. None of the central buildings was visible from the air, as they were all below tree level and the branches of the forest covered them completely. There were at least forty secret weapons establishments in this one unit, most of them devoted to the improvement of armour and the testing of ballistic projectiles. A large supersonic wind tunnel was built, and – for topographical reasons – the air intake had to be on open ground. So the German specialists erected a dummy farm-house to occupy the site, complete in every detail; and on one end (where the air intakes were) was a small out-house. Its roof slid sideways in its entirety to reveal the jet ducts when the device was going to be in use, and then they were quietly and unobtrusively slid back again afterwards, leaving the supporting beams standing rather conspicuously alongside. But no-one ever noticed.

And so it was that this immense establishment was erected and kept in full operation throughout the war without anyone knowing about it; two bombs did fall near the site during the entire war, but they were errors on bombing raids aimed at the town nearby.

At Ruit, eight miles or so from Stuttgart, another such institute (also named after a leading aviation leader) was established, the *Luftfahrtforschungsangstalt Graf Zeppelin;* but this had more of the traditional appearance of a German research

An early test rocket motor

The devastation at Peenemunde after the first heavy Allied raid during the night of 18th August 1943

One of the main streets of Peenemünde after the first raid

centre. As such it was soon located by Allied Intelligence, and bombed.

This institute was basically concerned with the then new science of aerodynamics. Models of secret weapons – rockets, missiles, and so on – were tested under extremely sophisticated conditions.

At Peenemünde an immense establishment was erected at a cost of over $120,000,000 (£50,000,000) to house, eventually, over 2,000 scientists. They were there to study rocketry, and particularly to build the A-series which gave rise to the V-2 (or A-4, as it was known to the scientists). The centre was built on an island at the mouth of the Oder, now the border between East Germany and Poland, but at the time still in Germany itself. The island is called Usedom and to fly over the area today, as I have recently done, demonstrates how unlikely it was that the British reconnaissance authorities would ever show much initial interest in the site as a centre for top-level secret developments. It was too far out from the centre of things: too much out on the limb. And the scattered buildings that did show up on routine pictures were quite typical of settlements dotted all over the German countryside. But this was where much of the most revolutionary of all the secret weapon development was centred. At the far north of the small island were the main test area and launching pads; along the coast lay the production plants and at the south of this stretch were the personal quarters of the staff. Behind this area were the barracks housing the military in the region.

Some almost routine bombing was carried out in 1943, when much of the area was shattered; but the main guidance control systems building – where much of the most vital research was going on – escaped undamaged. Even so, over 800 of the people on the island were killed when the raid took place, in the middle of August. After this, it was realised that some of the facility had better be dispersed throughout Germany; thus the theoretical development facility was moved to Garmisch-Partenkirchen, development went to Nordhausen and Bleicherode, and the main wind-tunnel and ancillary equipment went down to Kochel, some twenty-four miles south of Munich. This was christened *Wasserbau Versuchsanstalt* Kochelsee (experimental waterworks project) and gave rise to the most thorough research centre for long-range rocket development that, at the time, could have been envisaged.

They built a wind tunnel in which the air speed could be raised to the order of 3,000 mph, far better than anything else envisaged elsewhere in the world at that time. To many scientists the very idea of such an air velocity would have seemed impracticable without a vast fan unit to propel it: but the Kochel team designed instead a system which made the atmospheric pressure do the work for them. They constructed a vast pressure vessel of nearly 10,000 cubic feet and equipped it with a fairly powerful exhausting pump. In this way it could be reduced to near-vacuum in a very short while. At the moment that the test was to take place, a valve was opened admitting the atmosphere through an experimental chamber one and a half feet across and the model projectile inside was photographed during a whole range of air speeds, to show exactly how it would behave; and small pressure tubes were situated all over the models, flush with the surface, to measure the pressure changes produced by supersonic flight. The results were not perfect in some respects (for instance, there were problems of erosion of the chamber by the high-velocity air-flow, and – because it was working in a partial vacuum – the chamber was always below air pressure and this in itself introduced discrepancies of a minor order).

The Kochel apparatus was, then, a supreme example of advanced apparatus; yet in one respect at least it suffered from a fault often found in German war-time secret research. This was a simple lack of effort in the field of making instruments for taking experimental readings: the pressure tubes, for example, ran to small u-tubes filled with fluid. During a test, a dozen or so technicians would cluster around, all taking notes feverishly and memorising what took place. At no time, apparently, did anyone make an automatic plotter to do the job mechanically so that the recorded results, drawn on a roll of paper, could be examined later; indeed no-one even thought of taking photographs of the tubes for examination and accurate interpretation afterwards.

This failure to provide good instrumentation for experimental work is often clear from a perusal of the reports of the time. However this did not apply to the apparatus for the test itself, which was always of a high quality. The shock-wave photographs at Kochel, for example, were taken by the most sophisticated apparatus specially developed by companies such as the Zeiss organisation.

So good were the results that the Germans envisaged an even better tunnel, with a peak air velocity of 8,000 mph; they were going to construct a tunnel through more than a mile of rock to an industrial reservoir

Major-General Dörnberger (right), head of rocket research development, leads Field-Marshal Keitel (left) to inspect his handiwork

several hundreds of feet higher than the establishment itself; the water pressure, they felt, would drive high-speed turbines and produce a positive air flow of the order required. But this tunnel was never built before the war came to its end.

Even more grandiose in some respects was a gigantic tunnel, twenty-five feet across, capable of working at up to the speed of sound which was under construction at Otztal, Bavaria, when the war ended. Here too turbines driven by falling water from a nearby source were to have been the motive force for its operation.

Much useful work was done in ballistics at the *Technische Akademie der Luftwaffe* – the technical academy – under Schardin, one of the leading ballistics experts of the time. There were altogether thirteen institutes in the *Akademie*, covering subjects as diverse as physical and mechanical sciences, aircraft performance and control, and the performance of engines. It also carried out much definitive work on the functioning of explosives in shaped charges: depending on whether the charge is flat, spherical, or concave, the effect of the blast of a given amount of contact explosive can vary enormously – this is how it is that the slow, ponderous shell of a bazooka can blast a hole through the armour of a heavy tank.

This, then, was where the research was done. The conditions and pay were excellent, morale was high, and results were widely acclaimed. Not only that, but the deployment of this varied, vast conglomeration of facilities was intelligently done in view of the war situation, and the ingenious camouflage employed for many of them, the false buildings and sliding roofs, kept their work and even their existence a complete secret – not only to the Allies, but indeed even to the Germans themselves. Such a set-up is ideal for the furtherance of secret work, and the German secret weapon programme pressed steadily ahead as a result with incredible and in some cases devastating results.

One of Germany's leading rocket scientists, Dr Thiel (centre), briefs his staff in front of an early V-2 at Peenemünde

The magic touch

Three experimental anti-aircraft shells developed during the latter part of the war. *Left to right:* 128/105-mm discarding sabot shell; 105-mm ram jet shell; and 105/88-mm squeezebore shell. None of them entered general use

Many of the most secret of all Germany's weapons developed during the war were also the most bizarre; from devices to spot infra-red 'snipers' in the dark to sound rays that would kill, the range was considerable. Many of the ideas did not come to fruition, of course, and many of those that were produced failed in some practical respect. But in many ways they were to point the way to future developments in post-war scientific research.

Some of these reported developments were almost certainly mythical; the result of a romanticised picture of high-pressure research when viewed retrospectively. Thus the 'sun-gun' – which was reputed to be able to concentrate the sun's rays and literally burn an aeroplane out of the sky – is pure invention. But the equally unlikely sounding vortex-gun, producing an artificial whirlwind, is quite genuine.

This device was built and tested by a Dr Zippermeyer at Lofer, an experimental institute in the Tyrol. It consisted of a large-calibre mortar barrel sunk in the ground; and this fired shells containing coal-dust and a slow-burning explosive. The effect of this was intended to be enough to create an artificial whirlwind which would bring an aircraft down out of control. Would it work? Undoubtedly it could, given the most favourable circumstances. A large number of high-speed films for analysis were taken, and they showed that the forward-moving explosion of the coal dust was indeed sufficient to initiate the formation of a sizeable vortex. No-one knows whether the pressure changes would have been enough to cause frame failure of an aeroplane attacked in this manner, but the stresses on the wing loading might well prove to be excessive: clear-air turbulence has been known, in recent years, to bring down large airliners in pieces and it is feasible that Dr Zippermeyer's unlikely-sounding cannon could have done so too. However the device was never used in practice (although even the prototype might have had an effective range of perhaps a hundred yards or so). However similarly constructed shells which initiated a powerful, spreading methane explosion were used in Warsaw against the

Polish freedom fighters towards the end of the war.

The 'wind gun' was also realised in practical form. This was a strange device, a large angled barrel with a crooked elbow resting in an immense cradle like some enormous broken pea-shooter lying askew. It worked by the detonation of critical mixtures of hydrogen and oxygen in molecular proportions, as near as possible; the violent explosion triggered off a rapidly-ejected projectile of compressed air and water vapour which, like a solid 'plug' of air, was as effective as a small shell.

At least in theory! Practical tests of this strange device at Hillersleben, on the testing field, showed that planks of wood roughly one inch thick could be broken at a range of about two hundred yards. Here, nitrogen peroxide was used in some of the tests in order that the brown colour of the gas would enable the path and destination of the otherwise invisible projectile to be observed and photographed. The tests certainly showed that a very powerful zone of compressed, high-velocity air could be developed with sufficient force to cause some limited damage; the aerodynamics of a flying aircraft, however, would almost certainly have eliminated the effects of this weapon altogether. It was in fact installed on a bridge over the Elbe, but whether there were no aircraft or simply (as one suspects) no successes, the fact remains that the 'wind gun' was an interesting experiment – but a practical failure.

Invisible, dangerous air pressures were developed in another type of weapon designed at Lofer. This was the so-called 'sound cannon', and it was designed by Dr Richard Wallauschek. It took the form of large paraboloid reflectors, the final one in the series being over ten feet across, which were connected to a chamber made of several sub-unit firing tubes.

The purpose of these was to allow an admixture of methane and oxygen into the combustion chamber, where

Three experimental projectiles. *Left to right:* fin-stabilised 150-mm shell; shell for 310-mm gun; 40-mm arrow projectile

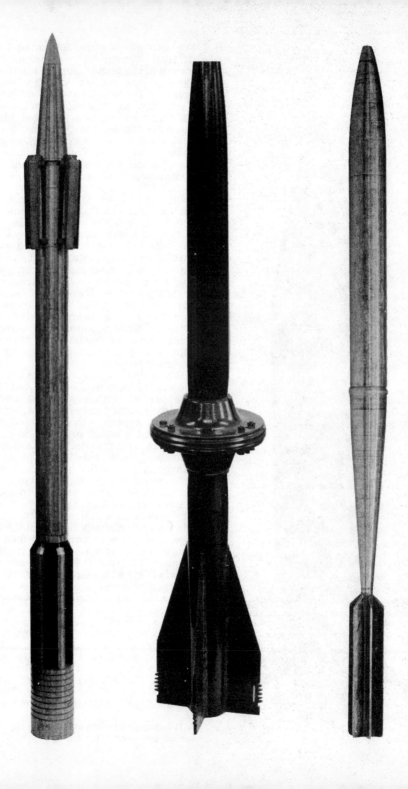

the two gases were detonated in a cyclical, continuous explosion. The length of the firing chamber itself was exactly one-quarter of the wavelength of the sound waves produced by the continuing explosions, each one of which, by producing a reflected, high-intensity shock-wave, initiated the next, so setting up a very high amplitude sound beam. This high note, of unendurable intensity, was radiated at pressures in excess of 1,000 millibars some fifty yards away, above the limits that man can withstand. At such a range, half a minute would be enough to kill the average man and at longer ranges, say, up to 250 yards, the effect would be excruciatingly painful and the result would be to incapacitate a soldier for some considerable time afterwards. No operational or physiological tests were carried out, although it has been suggested that there were several experiments using laboratory animals to demonstrate the basic feasibility of the concept. The 'gun' was never used for its designed purpose.

It has long been the dream of man to see successfully in the dark. And this was one of Germany's most enterprising 'magic gadgets' – a small, portable viewer which enabled the user to have effective vision in total darkness. It consisted of a hand-held chamber which acted as a picture-converter, changing invisible infra-red rays into visible light. At the front was a convex lens designed to focus the rays onto a light-sensitive screen. These rays were converted into cathode rays by this middle part of the device, and these in turn were focused on to a fluorescent screen. In this manner infra-red radiation was rendered visible as though it was showing on a portable television tube. At first it was used on its own, as a locator of infra-red emissions. Thus it would show a bright spot when pointed directly at a source of infra-red rays and therefore a hidden user of the radiation would easily be located and could be fired on. But before long the German troops were equipped with infra-red radiators themselves. These enabled them to illuminate the scene

A 150-mm shell with stabilising fins which emerged after firing

with invisible beams which, on reflection, were converted by the 'magic eye' into a visible picture.

Infra-red detectors were able to detect gunfire at a range of over eighty miles with an accuracy of one minute of arc, but the viewers used by individual troops were equally responsive to direction, so the process of sighting and aiming of a gun in total darkness became remarkably similar to that involved in normal daylight. It is, on the physical side, interesting to note that there is an apparent anomaly in the construction and mode of operation of such a detector. Infra-red radiation is of a lower energy level than visible light, which seems to make image-conversion impossible. The ultra-violet converter, for example, works by absorbing ultra-violet on to a phosphor from whence it is re-emitted at a lower energy level and therefore can be seen by the naked eye. Infra-red, however, is already lower in energy – how therefore can it be made to emit light rays of a higher value?

The answer was, in the case of the device already described, to feed in high-tension electricity across the cathode, which caused the rise in energy level. But in the hand-held (or gun-sight mounted) detector, the phosphors were specially selected to absorb energy from sunlight and store it; the effects of the infra-red were to cause the release of the stored energy from the phosphors. And so in this way an apparently basic rule of physics was neatly circumvented, giving rise to a most valuable secret device which the Germans valued highly. It was amongst their most guarded military secrets. However, it did need to be 're-charged' by sunlight and so unless it was left in bright sunlight for perhaps a quarter of an hour each day it lost its potency, and even then it had a limited life without a boost of solar radiation.

But apart from detecting the enemy's position, the German scientists threw up many ideas for confusing enemy observations. On one hand they were reported to have built automatic meteorological stations which

Rocket shell for the massive 280-mm railway gun

Above: Part of the ground equipment used during the development of the V-2. *Left:* The nose fuze, base fuze, and exploder of the V-2 rocket. *Top right:* Destruction inside the main assembly area for V-2s at Peenemünde following the first heavy Allied air raid. *Bottom right:* Test firing a recoilless anti-tank gun

German ballistics devoted much research to shells which could be fired in 'bundles'. Above are some of their developments

transmitted information back to Germany at periodic .intervals, and which were left in the Atlantic for remote operation. And on the other hand they produced a series of experimental dummy submarines; these were loaded with high explosive and the intention was that Allied vessels would ram the dummy and so sink. But there are no reliable rumours of the idea catching on with the officials at ministry level.

A weird remote-control 'boat' – the Tornado – was made by strapping two seaplane floats to a V-1 pulse-jet engine and fitting it out with a 1,200 lb bomb charge. It was intended to skim over the surface of the sea and explode violently against the target. In fact it never made more than 40 mph and it capsized in all but fair weather, and never proved to be a success. Some fifteen foot motor-boats (powered by American Ford V-8 engines) were also produced. They too were guided by remote control, or set on course by a pilot who jumped overboard when the target was near, and

contained a 1,000 lb bomb on board. However they were not very successful in practice and they, too, were later discarded.

But many of the weapons which have been discussed more openly in the postwar years (and which were used as propaganda threats in the war years too) simply did not exist. There was no super-virus on the way to wipe out vast areas of enemy territory, neither were there any 'death-ray' developments (with the dubious exception of the sound-cannon described earlier).

There was reported to be interest shown in a bomb in which the thermal properties of its reaction were endothermic – i.e. it would generate intense cold on detonation, rather than fierce heat. The idea was said to be the freezing of vast areas – perhaps a mile radius from the bomb – and the rapid removal of all life from the affected region which would soon return otherwise to normal. But this did not exist either, other than as a high-flown and totally impracticable idea. There are supposed to have been 'flying saucers' too, which were near the final stages of development, and indeed it may be that some progress was made towards the construction of a small disc-like aircraft, but the results were des-

troyed, apparently before they fell into enemy hands. However the fanciful reports of some writers are pure invention: among them are accounts of faster-than-sound flying saucers which could rise to altitudes of perhaps 40,000 feet within a few minutes. It would be interesting to assume that these were true, and that some of the sightings of unidentified flying objects might be due to modern experimental 'saucers' built with postwar knowhow. Indeed the theory has been advanced on several occasions – but it is pure inventive speculation.

And finally, as an example of the zanier German secrets, we may cite the cable bomb which was envisaged as an ultimate answer to enemy bombing raids over German soil. These bombs were to have been dropped in front of an advancing horde of aircraft, each bomb unreeling a cable as it fell away at speed. The upper end of the cable was kept in the air by either a balloon (the first suggestion – although a balloon with adequate lifting capacity would have been almost as large as the aircraft delivering the weapon!) or, it was suggested, a parachute. In this way the enemy aircraft would fly unwittingly into a snare of aerial wires which

would entangle in the propellers and wind rapidly around them, lifting the bomb up into the air until it detonated on the side of the bomber's engine housing and sent it down out of control – another victim, it was planned, of Germany's top-secret super weapons. In practice they would not have been successful, since the inertia of the spinning propeller would probably snap a cable which it encountered at speed and the number of parachute bombs that would have been necessary to intercept even by simple mechanical means would have been prohibitive. There are some records of successes in action, but they seem, like several other of these 'guarded secrets', to have gained a good deal in the retelling of the story after the event.

But some of the other secret weapons were not just talk: they were not mere experimental freaks. Many of them, as we shall see, were important, indeed earth-shattering developments in every sense of the word – and still they are marvelled at, as objects built by a fanatical and obsessive nation under conditions of misplaced zeal and a tragically, cruelly misguided conscience.

35

Germany's terror weapons

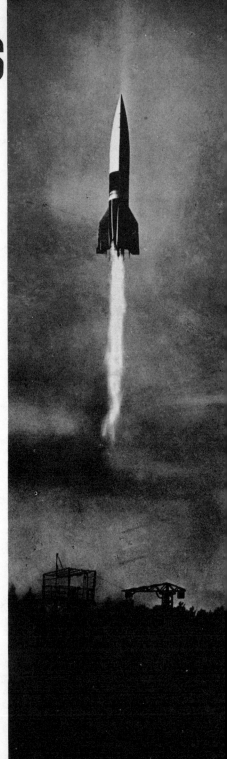

The triumph of German rocketry, a V-2 soars into the air

It was the rocket which gave Germany her main hope of building – as she eventually did build – the most devastating weapons of their time. To this day it is the ballistic missile which is at the root of the modern balance of power, and the entire heritage of the present-day rocket system is a direct derivation from the developments in Germany in the war years.

But there is more to it than that: the Germans' work in this field virtually laid the foundations of the whole subject; the very first liquid-fuelled rocket in history flew only thirteen years before the outbreak of hostilities in 1939, and so it was entirely during that period that the rocket changed from being a dangerous toy into a most sophisticated weapon of war. Thus the history of German rocketry is in essence the entire story, from the very beginnings of experimentation.

The start of theoretical interest in the subject came from three workers, a Russian, an American, and a German. The Russian, Tsiolkovsky, came into the field quite cold. There was no earlier thinker on the subject at all, apart from a political prisoner named Kibalchich who wrote a brief account of a space-craft propelled by charges of gunpowder – but he was executed for offences involved in an assassination attempt against the Tsar before his ideas were ever expounded in any kind of even preliminary detail. It was really just a passing whim, a fancy; but Tsiolkovsky took it from that and developed the theory of rocketry into a thoroughly worked-out concept, which he published as a paper in a Russian scientific journal named *Naootchnoye Obozreniye* in 1903. He talked in this paper of the effects of a rocket powered by liquid-oxygen/liquid-hydrogen – one of the most successful of today's combinations. He even went so far as to write of a multi-stage rocket, 'the people's space rocket train' of the year 2017, which may not have been a bad estimate.

After the revolution, Tsiolkovsky continued to gain in favour and was elected in 1919 to membership of the Socialist Science Academy. But the Russians did not appreciate the implications of his detailed day-dreams about planetary travel; they thought of him as a visionary, rather than an innovator. And after him there were no great Russian rocket experts of military significance.

It was the American Goddard who trod the path towards practical realisation of the aim. It was he who made the practical models and tested them experimentally, and it was his rocket which eventually became the first liquid-fuelled rocket ever to fly. Robert H Goddard was before his time: he was never recognised as he ought to have been, indeed it was in 1959 that the first section of an autobiography of his appeared in print. It had been written in 1927, and he died in 1945, just a few months after the war had shown how far-reaching his earlier work had proved to be.

Even before receiving his PhD in 1911, Goddard had been working on liquid-fuelled rocket theory. Then, in the First World War, he developed two pioneer rockets which could have proved successful in the field – but which the ending of the war eclipsed altogether. The following year he applied for a research grant to continue his work; the affidavit in support of the application, entitled 'A method of reaching extreme altitude', is the first of his publications on the subject in any depth. His first rocket flight – when a spindly contraption remained aloft for less than three seconds, launched from a six-foot ramp – was recorded on 16th March 1926 at Auburn, Massachusetts. In the following month he broke his record by getting a model to stay up for over four seconds; within three years he had successfully carried out flights with small rockets that exceeded 60 mph over distances of perhaps 200 feet and the following year, 1930, saw a 2,000 foot altitude test at 500 mph. Though Goddard died largely unrecognised by the informed world at large, he had already laid the foundations of a secure and unique place in history.

But it was the German research worker who, following on the heels of these two men, dragged rocketry out of the realms of gadgetry and into the realm of science-fact.

The great German pioneer was Hermann Oberth; he read Jules Verne with passion as a youth, and later the

Wernher von Braun (in dark suit) with a group of high ranking officers. Lieutenant-General Schneider, head of the Army Testing Office, is next to von Braun

Above: Major-General Dörnberger with Professor Oberth in 1941. *Below:* Dörnberger shows Heinrich Himmler round Peenemünde in 1943

Above: Field-Marshal Keitel watches intently through his field-glasses.
Below: Dr Todt, Economics Minister of the Reich (second from left)

Johannes Winkler, who first sent up a camera in a rocket in 1931

works of Tsiolkovsky and Goddard. Indeed his first work on rocketry, which was published as a ninety-two page booklet when he was twenty-nine, contained many of Goddard's ideas. That was in 1923. A year earlier he had written to Goddard and asked him for some of his reprints – but the 1923 booklet under Oberth's name (entitled 'The rocket going into interplanetary space') had a long explanation of the coincidental nature of his work, emphasising that he had not plagiarised Goddard's work in any way.

The unsympathetic have sometimes inferred that this disclaimer was insincere. But was it? The great quality of Oberth's work is such as to refute the notion: it would clearly have been unnecessary for him to plagiarise. Quite possibly Goddard's work, and that of the Russian, were of great value in encouraging Oberth in his endeavours, but the results he later attained were far more prestigious. Indeed, in this first pamphlet he drew a rough section through an instantly recognisable rocket, not unlike the V-2 at first glance. This was the Model B which, though never built, was a clear foretaste of what lay in store.

Oberth claims to have proposed as early as 1917 to the German war ministry that liquid-fuelled long-range rockets could be used as weapons of war. The idea was turned down. But Oberth continued to try to press the notion of such devices until they were accepted: eventually he was, in the late 1920s, appointed to work on a space-fiction film entitled 'The Girl in the Moon' and commissioned to build and fly a rocket for the effects department of the film production company. But it came to nothing; one or two static firings of a motor were carried out, but the building of the rocket was a complete failure. None the less, the very fact that a film was being made of that subject at all showed the degree of public interest in it – and Hermann Oberth, though he probably did not realise it at the time, had through his much-publicised exploits done much to condition the minds of militarists towards the eventual use of the rocket in war. Within ten short years he was to see how very necessary such an idea would become to the

German war effort.

In March 1931 a scientist named Karl Poggensee is reported to have flown a solid-fuel rocket to a height of well over 1,500 feet carrying a parachute, altimeter device, and a camera to record what was happening. In the same month Johann Winkler and Hugo Hückel flew their Mark 1 rocket to a similar altitude – and this one was fuelled by liquid methane and liquid oxygen, a difficult blend to handle at the best of times. Later that year the great German pioneer Willy Ley flew a small, stubby rocket fuelled with benzene and liquid oxygen to heights eventually up to a mile. Rocketry, at least as a feat of inventive exhibitionism, was clearly in Germany to stay.

But then came the slump, and not much could be done to provide funds for rocket societies and amateur experimenters. The *Racketenflugplatz* or 'rocket-flight field' which some of the amateur enthusiasts had been using on the outskirts of Berlin was reconverted back into its original purpose, an arms dump, and the many rocket hobbyists who had done so much to arouse interest in the subject disbanded. Meanwhile the Army Testing Office continued to develop its proof and test establishment at Kummersdorf for the purposes of rocket experimentation. In charge was Captain Walter Dörnberger, later to become a General, and it was he who led the secret rocket weapons development from that moment on.

Dörnberger's brief was simple: to develop, design, and construct topsecret weapons of a hitherto undreamed-of nature – to turn out high-velocity rockets, rocket-assisted torpedos, anything in fact which could possibly give Germany an edge over the enemy when hostilities began. But the Hitler machine, though it wanted results, was not over-willing to pay for them; the rocket programme was therefore limited in its scope from the first. There was little room for sudden expansion, little opportunity for startling research.

But there is one factor that even limited finance need not prevent, and that is the selection of able staff. Dörnberger could not take on many people when he started work on the rocketry projects for the government, but he did encourage one young student to work for a PhD at the centre. This youth, a self-assured almost chubby youngster in his very early twenties, had sometimes helped with rocket experiments at the *Racketenflugplatz* and was wildly enthusiastic about rocketry. He was given a space in which to work and a course of study, and allowed to have a mechanic to help him with construction work of an experimental nature. Dörnberger felt that the boy had a promising future: his name was Wernher von Braun.

Reading some recent accounts of the activities at Kummersdorf creates a very exciting impression of activity and creative freedom. In fact it was not as exciting for everyone there. Due to somewhat inefficient leadership – at least in the public relations field – Waffenforschungs never managed to establish itself as the leading authority on secret weapons in these important early days of the German effort. But Captain Dörnberger's men were fanatical about their subject, and soon it was the rocket laboratory that came to prominence at Kummersdorf. Dörnberger was later promoted to Major-General and from this greater position of eminence he proceded to gradually expand and enlarge the whole complex. War was now in the air, and from 1932 to 1936 the staff increased to about sixty; by the outbreak of war it was nearly 300, the cream of Germany's technicians and rocketry scientists. They were a new breed – and so were the weapons they produced.

The first rocket to be produced there was destined to be a direct forerunner of the V-2. Code-named Aggregate-1, the rocket was a small 660-lb thrust device. Fuelled with alcohol and liquid oxygen – both forced into the combustion chamber by pressure from a liquid-nitrogen tank – the rocket motor fired successfully in static tests. The rocket itself, however, was not as successful. Stabilised with a gyroscope in the nose, it was too hazardous to use since the fuel mixture, a highly inflammable one, tended to explode violently. In addition it was later calculated that the gyroscope was in the wrong place, and so the A-1 was scrapped.

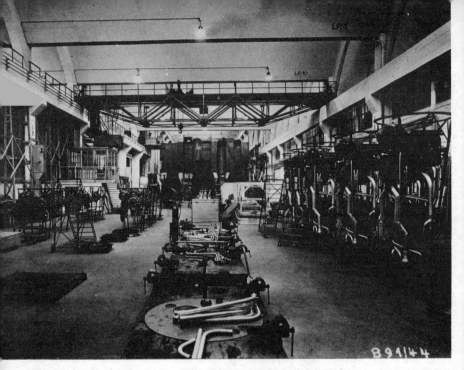

The assembly room for V-2s at the test centre in Peenemünde

Work was then started on the A-2. And this was more practical by far: two of the rockets, named Max und Moritz, after two German comic-book characters (they still feature in some German comics at least!) were sent on vertical test flights to altitudes of over 6,000 feet. These tests took place on the island of Borkum, off the German North Sea coast, formerly a well-known resort area and situated near the mouth of the River Ems. The peerless success of the firing came at just the right moment. It was the beginning of 1935, and the pace of the secret weapon development was beginning to quicken. Dörnberger was allotted more staff and a bigger budget, and with them larger and larger rocket engines were designed.

However it soon became apparent that facilities were now too cramped and so the organisation established a group centre at Peenemünde, the Baltic island which we visited in the first chapter and which proved to be so vital towards the whole war effort in its later stages. This soon became the centre of activities.

In the spring of 1937 the Peenemünde establishment was first taken over by the group: and as early as September of the same year the first of the test rocket firings was held. This was of an enlarged, improved, modernised rocket dubbed the A-3 and developing over 3,000 lbs of thrust from its liquid-oxygen/alcohol engine. Each rocket weighed 1,650 lbs and stood twenty-one feet tall; supported in its test stand of steel girders the static tests were reported to have been a most impressive sight. It was an historic sight, too, for this was the first large operational rocket to have been envisaged.

But still it wasn't good enough. The guidance system of the rocket was too primitive and susceptible to failure, and the A-3 never flew successfully. But by now Germany was alight to the need for long-range secret weapons and a whole range of companies in the private enterprise sphere became involved with varying aspects of the research. The Luftwaffe invited Austria's leading rocketry scientist, Dr Eugen Sänger, to set up a laboratory at Trauen where he succeeded in producing liquid rocket motors, fuelled by diesel-oil and compressed oxygen,

An early V-2 is brought out of its camouflaged lair

which would run for half an hour – then an incredible feat.

Such was the degree of interest in the development of weapons like this, that Dr Steinhoff, one of the leading rocket scientists at Peenemünde, is quite certain that perhaps one-third of all Germany's scientists during the early war years must have worked on the long-range rockets at one time or another and in some capacity – many of them, no doubt, without knowing what was the fate of their own effort; as far as they were concerned, they were simply carrying out a commissioned investigation into navigation control systems, tele-communications, fuel pumps and so forth.

And so, at Peenemünde, the research effort was stepped up. As war drew closer the War Ministry sent a demand for the 'ultimate weapon': it must be a rocket capable of delivering a ton of high-explosive to London, perhaps, or Paris; but in any event with a range of over 150 miles. It must be undetectable and completely reliable, and able to overcome all attempts at counter-attack by the enemy. A tall order!

The results were fairly obvious – it had to be a new monster rocket, yet not so large as to be difficult to transport. It was probable that any large and conspicuous launching site might be attacked by bombs or – who knew then? – the Allies' own monster rockets, and so it was necessary to bear in mind the need for the rocket to be transportable to semi-portable launching sites elsewhere. It had to fit the railway tunnels of Germany at that time, for instance, for transportation by train; it had to use readily-available materials in case of blockade difficulties; it had to be capable of mass production – and, above all, it must be reliable.

And here was trouble! For the A-3, the best rocket then developed, was a near-uncontrollable mechanical zombie with plenty of temperament. It needed a new, foolproof guidance system and control equipment – and it needed power. To develop the rocket was now a matter of urgency and so the A-4 began to take shape.

At least it did in theory. It soon became apparent that there were many important problems to overcome, and so it was decided to make

Left: The lower part of the motor of a V-2 showing the circular hydrogen peroxide tank and the firing chamber. *Above:* The motor under test

the first tests on a small-scale model of the A-4. This, the A-5, was virtually the same size as the earlier A-3, but it had a simplified guidance system, entirely new controls, and a greatly modified system of engineering construction. Late in 1938, it was fired out across the Baltic to a height of over 35,000 feet. In the next year or so, thirty more were tested, many with subsequent parachute recovery. But the feasibility of the idea had been demonstrated successfully by then.

Work on servo-control systems and high-capacity fuel pumps inevitably led to the building of the first A-4 – which promptly failed when tested. On 13th June 1942 this pioneer rocket, the first V-2, was checked, rechecked, and pronounced ready for firing. It stood 46 feet 1½ inches tall, weighed 27,000 lbs and was fuelled with methyl alcohol and liquid oxygen. The pumps were started, ignition achieved, and the rocket rose unsteadily from its launch pad. In a billowing cloud of smoke and steam it

rose, gaining speed, and then – at just the wrong moment – the propellant pump-motor failed. The rocket staggered a little higher, then fell sideways and crashed, sending up clouds of misty fumes from the ruptured fuel and oxygen tanks. On the 16th August the second A-4 was fired, with more success, and although the motor failed prematurely, possibly for the same reason as the first, telemetry indicated that the device had exceeded the speed of sound. That too was a moment of history.

But the third was a complete success. On 3rd October 1942, this A-4 was fired out along the coast of Pomerania; the engine burned for a minute or so, boosting the altitude to some fifty miles and it fell to earth 119.3 miles away – the age of the missile had arrived.

Suddenly the German government was interested. Dörnberger had struggled for years to gain recognition for his group's potential, but with only limited success; Hitler had even been

Lehrmeder

„Erika"

A unique photograph of the A-5, the small-scale model of what was to become
the V-2, being tested underneath a Heinkel 111 in 1939

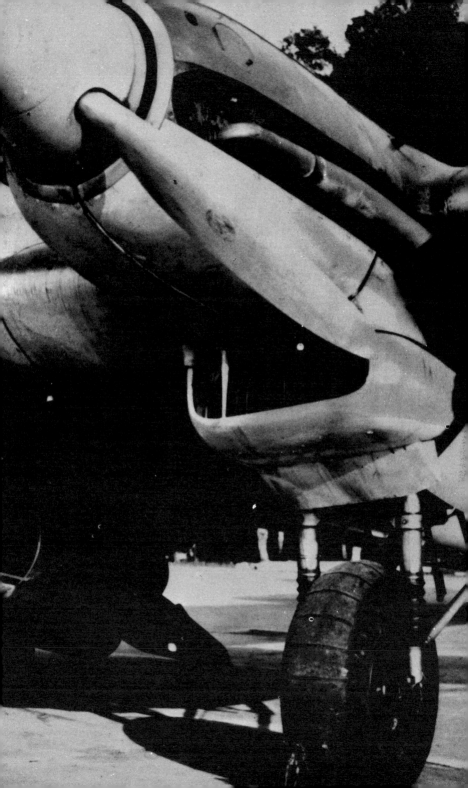

Heizbehälter: Betriebszustand
(theor. Werte)

Above: A German diagram showing the firing-chamber of the V-2, and charts recording firing data. *Right:* A cutaway of the V-2 showing the large oxygen and alcohol tanks above the motor. *Far right:* The V-2 on its launch platform showing its size.

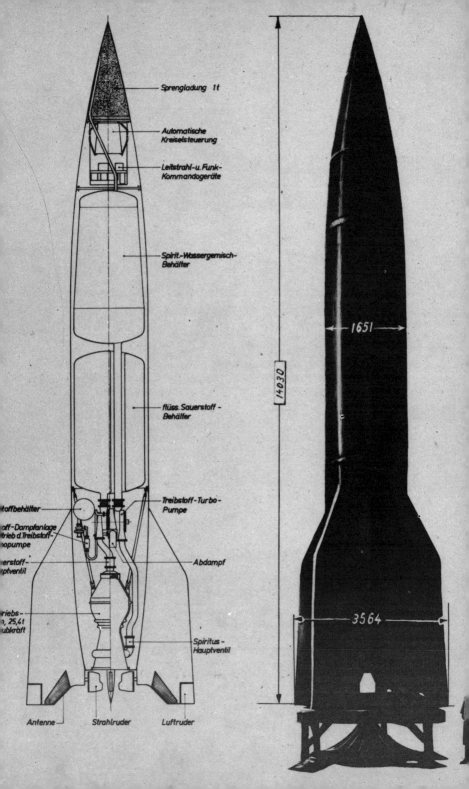

Sprengladung 1 t

Automatische
Kreiselsteuerung

Leitstrahl-u.Funk-
Kommandogeräte

Spirit.-Wassergemisch-
Behälter

flüss. Sauerstoff-
Behälter

Treibstoff-Turbo-
Pumpe

...stoffbehälter

...off-Dampfanlage
...trieb d.Treibstoff-
...nopumpe

...erstoff-
...uptventil

Abdampf

...riebs-
..., 25,4 t
...ubkraft

Spiritus-
Hauptventil

Antenne

Strahlruder

Luftruder

1651

14030

3564

Above left: The mobile extending tower used to work on the warhead and guidance mechanism of the V-2 when in the firing position. *Left:* Wheeling fuel out to the A-3 test rocket during experiments in 1937. *Above:* The trailer for the V-2 which could be used to raise the rocket into the firing position. *Right:* Technicians work on a shrouded A-3 rocket during tests at a secret site on the north coast of Germany

wooed to come and watch static firings at Kummersdorf. He was not greatly impressed by the fire and smoke. But the solid, undeniable achievement of such a revolutionary new weapon was a different matter altogether, and he at once appointed a special V-2 production committee to help and coordinate developments, under the direction of General Degenkolb; but it was less a help than a hindrance and von Braun is reported as having described it tersely as 'a thorn in our flesh'.

However more money, increased staff and better equipment were put at the disposal of the German scientists at Peenemünde, and the production of V-2s was stepped up – not initially for tactical deployment, but for the purpose of training troops and technicians and gaining experience in the handling of the weapon. Eventually modifications were incorporated to increase the rocket's range to 260 miles, and its speed to over 3,300 mph. Many independent organisations and companies assisted with research and development, among them Zeppelin Luftschiffbau and the Heinkel works in the Tyrol; but eventually the final version of the V-2 – the one which went into production and of which over 5,000 were produced – was a most impressive weapon, and certainly a magnificent rocket for its time. It was 46 feet tall, 5 feet 5 inches across, and weighed over 12½ tons at launch, 70 per cent of that being fuel. It carried 8,300 lbs of fuel, 11,000 lb of liquid oxygen which were consumed in the combustion chamber at the rate of 275 lbs per second. The exhaust-gas velocity was 6,950 ft/sec and the accuracy of aim was stated to be – in the opinion of the Germans – 'better than 4 per cent'. The secret of its controllability lay in a cybernetic servo-system which directed the vanes in the exhaust stream: they turned from side to side, slightly deflecting the path of the rocket thrust and producing lateral effects which altered its trajectory slightly. In this way the rocket was kept upright during launch, and at the correct time was steered over towards its target trajectory. There were elevator controls, too, in the tail fins – but these were of secondary importance.

Especially during the vital early stages of launch – when the rocket's speed was far too slow to give any aerodynamic usefulness to the fin elevators – these rocket exhaust deflectors were the key to success. They were really the development which made for the success of the rocket as a whole; the prospect of it, or its one-ton warhead, being sent into an uncontrollable off-course trajectory and causing damage shortly after launch was clearly inadmissible. The concept was used by the Germans elsewhere as a means of rocket control, and it was used subsequently by others.

But at last Hitler was won over to the notion of long-range rocket bombardment of London, and he called for a detailed investigation of the practical implications involved. Hitler's idea seems to have been based on a massive onslaught of perhaps 5,000 rockets launched in rapid succession, or simultaneously as far as possible. But Dörnberger said no. There was every chance of making many thousands of rockets, he stated – assuming that the Führer made available great reserves of finance and materials – but the fuel position was not so good. Blockaded and largely isolated, Germany (in spite of all that was being said about her self-sufficiency and enterprise) simply could not spare such vast amounts of materials.

Throughout this time von Braun's enthusiasm for the potentiality of rocket travel was growing. At one stage, indeed, he was arrested by the Gestapo and put on a charge which alleged, in effect, that he was (a) not in favour of the bombardment of London and (b) secretly building plans for the use of rockets for the exploration of space, rather than for the military aims of the Fatherland. Dörnberger is reported to have insisted on his release, without effect at first; then he emphasised that von Braun was necessary for the further progress of the rocket programme. At this von Braun was set free.

Meanwhile Hitler was still toying with rival plans and ideas; he was keener on aeroplanes than he was on rockets, and rocket-powered aircraft, jets, and flying bombs were all nearer to his heart than the admittedly

Braun's notebook on rocket theory, written in 1929

successful monster rocket of Peenemünde. Speer was the first government minister to witness a rocket flight, and he was much impressed – indeed it was Speer's insistence on the merits of the idea which converted Hitler. But it was not until early in 1943 that Hitler was completely convinced. Peenemünde was finally given virtually unlimited financial support, Dörnberger unlimited power within his field, and as a matter of priority a giant factory at Friedrichshaven was taken over for production.

But of course there was a schematic anomaly in the whole affair. The rocket was regarded as armament: it was classified as a large shell which carried its charge within itself – in other words, it was an army responsibility. The Luftwaffe chiefs did not take to the idea at all; for all its contorted evolutionary channels it was still, they said, an aerial weapon and therefore it ought to be under the aegis of the air minister and not controlled by 'damned soldiers'. But they

soon had their own project: it became known as the V-1.

This was a largely wooden pilotless flying bomb. The development took place in the DFS – *Deutsche Forschungs-anstalt fur Segelflug* (German Research Centre for Gliders) – which was originally established at Darmstadt, former capital of the Grand Duchy of Hesse-Darmstadt and a well-known centre of carpet manufacturing. Later it moved to Ainring near Salzburg on the Austrian border. Georgii, who was one of the chief scientists at Fo Fü, was also leader of DFS, and was largely responsible for Germany's troop-carrying glider development. About 1,000 people worked there altogether, and several other secret weapons – including the *Wasserfall* project – were developed at the DFS.

The V-1 began its story in the 1920s, when a Munich professor, Paul Schmidt, began work on a pulse-jet device. How this worked is easy to understand : the moving duct, because of its forward motion, had air

Below: The development of the V-2 as an effective military weapon had its share of setbacks. *Right:* One rocket which failed to leave the launching pad

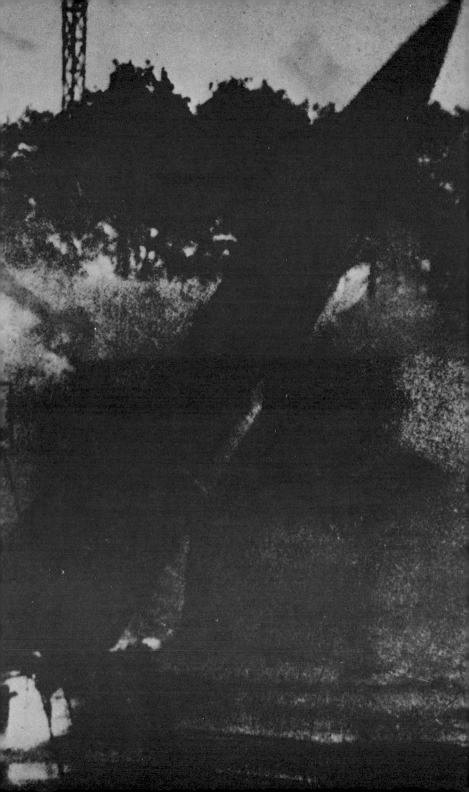

forced in through the hinged shutters at the front. Immediately behind them was a series of perforated fuel inlets and the draught of passing through air drew fuel out of the ducts and formed a fine aerosol which was ignited by a form of glow-plug. The near-critical mixture ignited and burned rapidly in the form of a dull explosion; thus the force of the detonation forced the shutters closed and the exhaust gases were driven out at the rear of the jet, forcing it forward. The continued burning of the fuel was at once prevented by the closing of the forward shutters and so the blast died rapidly away: once again the forward motion of the duct through the air caused the shutters to be re-opened, admitting another charge of fuel-laden air to the combustion chamber. In this way the cycle continued, and the device was driven forward by a regular series of low-frequency detonations. The device became known as the pulse-jet, or Schmidt duct, and as early as 1934 the professor was reported to have sug-

gested that the duct could be used to pilot a form of 'aerial torpedo' but the idea was turned down. Subsequently the duct was developed – as is often the case, without the inventor's involvement in the project – at the Argus Motorwerken Gesellschaft. Their product ran on gasoline and developed a thrust of over 700 lbs.

The short-winged, stubby airframe driven by the duct was designed and developed by Robert Lusser, chief designer with the Gerhard Fieseler Flugzeugbau; it was first known as the Fieseler Fi-103. And after much protraction and argument behind the scenes, in June 1941 the go-ahead was given for further development and production of the weapon in large numbers.

The technical specifications of the weapon were somewhat unpretentious; its speed, performance, and reliability were none too great – but it was cheap and easy to build, which was a very great advantage in itself.

It flew between 1,000 and 7,000 feet at

Left and below: The last frames of the film as the rocket crashes on its side and technicians examine the wreckage

Above: Another V-2 fails; a fraction of a second after firing the rocket bursts into flames and crashes to the ground. *Right:* Completed V-2s ready to leave the factory

a speed of over 400 mph; its range was initially stated to be 180 miles, later increased to about 250. It weighed about 4,800 lbs and of that nearly 2,000 lbs was trinitrotoluol and ammonium nitrate as warhead. It was 27 feet long and just under 5 feet in diameter.

The first tests, ironically enough in view of the interdepartmental rivalry which was rampant at the time, took place at the Peenemünde range, the only one well enough equipped with tracking apparatus, at the end of 1941. It proved to be a great success, and at once a controversy flared over the priority situation. The air arm was obviously keener to see the V-1 press ahead, whereas the army branch wanted to push their own V-2. What was to be done?

The merits of the two devices were worth comparing. The V-1 was cheap to make – it cost between 1,500 and 10,000 marks' (actual estimates vary considerably) – whereas the V-2 cost 75,000 marks. Each V-1 only consumed about 280 man-hours, according to one estimate, whereas the V-2 took 13,000. On the other hand the V-2 was supersonic and, although it had the same size payload, within approximate limits, its damage was conjectured to be greater because of its high-pressure shock wave produced by its speed of arrival – roughly Mach 4, or four times the speed of sound.

The V-1 was unjammable – it was not radio-guided, and so it was impossible for the Allies to find some means of deflecting it by interference with an electronic guidance system. But on the other hand the V-2 was faster by far – the almost lumbering speed of the V-1 by comparison meant it could be shot out of the sky or, as was done in many cases, tipped over by the wing of a fighter aircraft and left to plunge into the Channel.

It could be trapped by barrage balloon cables too, which the V-2 could avoid altogether; but on the other hand it was fuelled by cheap, low-grade petrol which could be distilled from Germany's native lignite beds whereas the V-2 needed alcohol and liquid oxygen.

The V-1 was unreliable, too; a quarter of them failed in use. The V-2 on the other hand was believed to be a world-beating missile, a record setter, an unprecedented weapon of war – and rightly so, for it was. Little wonder then that there was so much controversy about the relative merits of the two projects.

The solution was fairly obvious. Germany was fighting hard to overcome her opponents and to establish herself as a leading world power. In

The tail of a V-2 on its launching stand showing the symbol of the operating unit

An operational V-1 Flying bomb takes
off on its journey to London

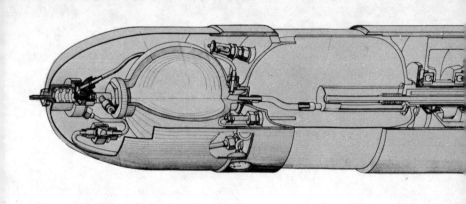

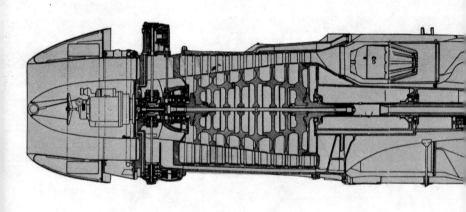

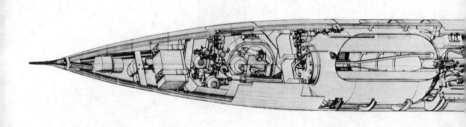

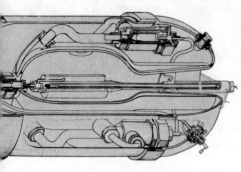

Three examples of the ingenuity of German rocket and aircraft scientists which were copied by Allied experts after the war.
Top: Diagram of the motor which powered the A-2 rocket of the early 1930s, a forerunner of the V-2.
Centre: An Allied drawing of the Jumo 004 engine which powered the early German jets.
Bottom: Detailed cutaway of the V-2, the forerunner of all the great rocket programmes

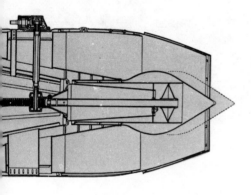

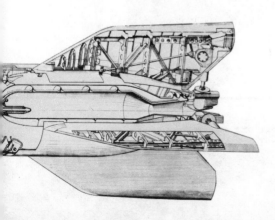

A V-1 is wheeled out of its protective hangar and prepared for firing

different ways she could do with both weapons: a cheap-and-cheerful version and a highly sophisticated rocket projectile. And so, after lengthy meetings and discussions, both projects were carried forward. As the British were to find out to their cost, both materialised.

From 1943 onwards, both projects were leading the German secret weapon drive. The V-1 was equipped with a simple but effective guidance system: it contained an Askania gyroscope for direction and attitude and on the nose there was a small propeller. This was driven by the flying bomb's passage through the atmosphere and though a system of cogs it activated a primitive distance recorder. At a preset distance the fuel was automatically tripped and the motor became silent; the whole contraption then fell in an oscillating dive to the ground and exploded. Launching of the V-1 proved to be difficult too: the duct will only operate, obviously, when the device is already moving forward at an appreciable speed. For this reason a number of conspicuous launching ramps were constructed on which the flying bombs were launched by a steam-driven device, giving them enough forward energy to open the duct flaps and commence the detonation cycle.

Later in the war, the 'Reichenberg', a piloted version of the V-1, was developed. It was not basically intended for suicide missions at all, as has been suggested by some commentators, but was intended as an experimental test-bed to iron out control difficulties. A higher-powered version of the V-1 with an increased range was envisaged towards the closing stages of the war, but it was never realised in practice.

Because of the growing fervour of Hitler and his aides to produce bigger, better, and more effective weapons, both the V-1 and the V-2 pressed ahead into routine production. The V-2 installations at Peenemünde were, as mentioned earlier, damaged by British action and as a result the construction was undertaken further afield. The wind-tunnel facility was moved out to Kochel, the research and design went to Garmisch-Partenkirchen, and the manufacture itself was taken to Nordhousen and Bliecherode. In the Harz mountains, subterranean workshops were established which proved to be invulnerable and, indeed, undetectable too. Thus the production of Germany's weapons of terror could continue apace; the safety of her aircrews was to be safeguarded, losses to heavy aircraft minimised – and the British were to be taught a lesson they would never forget.

Indeed, that proved to be very true.

Left: Retrieving the motor of the A-3 rocket after a test in 1937. *Above:* The wreckage of a crashed V-2. *Below:* An operational V-2

The secrets take wing

Natter takes off vertically from its launching tower

During their eternal search for new and better secret weapons with which to attack the enemy, the German scientists were peerless in the aero-dynamic field, and the incredibly varied range of new aircraft ideas which they threw up are still legend-ary. Much to the chagrin of the Ameri-can designers, they came to owe much to the techniques revealed by Opera-tion Paperclip, when Allied specialists moved across Germany and collected as much in the way of men and materi-als as it was possible to obtain.

German secret aircraft were many and varied; from small pilotless weapons – like the V-1 – to revolu-tionary flying-wing aeroplanes, their attainments were considerable. Much of it was due to the originality which they brought to the problems. And some of the results that were thrown up are adventurous and viable even by today's standards.

The *Natter* aircraft illustrates the broadminded, heterodox form of ap-proach to the problems of war. On 1st August 1944 the Chief of Development of the Reichsluftministerium – the Air Ministry – Herr Oberst Knemayer, was set a problem. Heavy bombing raids over Germany by Allied bombers necessitated a new means of approach. Was there something economical, fast, hard-hitting, and reliable that could be used to knock the enemy aircraft out of the sky?

It was a tall order – but one which set obvious criteria of performance. Herr Knemayer set out to list them and fulfil them as far as possible by criteria of design. And the result was the *Natter*. His design was simple: He would take a rocket-powered near-sonic aeroplane, fit it out with arma-ments, blast it into the path of the bombers, and then let the pilot bale out. He, and the aircraft, would be recovered subsequently.

The work commenced at Bachem-Werke in Waldsee late in 1944: by the end of the war 150 were on order for the SS, and fifty more for the Luft-waffe. The designers worked as a close-knit and expert team; with Knemayer were Bachem, formerly technical director of Fieseler, and Botheder, a Dutchman who had studied at Stutt-gart and came to work at Dornier in 1940. Subsequently the factory dis-persed, Bachem staying at Waldsee and Botheder being dispatched to send a group of four of the planes to St Leonhard – it was at this stage, in May 1945, that the Americans caught up with him and as a result of that the truth came out.

There had been 300-odd workers, sixty of them engineers, working on the *Natter* project, he said. Because of a dearth of materials and skilled labour, the device was designed for poor quality of raw materials and low-tolerance tool work; cheap grades of wood and the poorest sheet steel were used in construction.

By the end of the war there was a plan to sell the *Natter* to the Japanese, and they were also going to manufac-ture it under licence from the Ger-mans.

The remaining staff members of note, Botheder explained in garru-lous English, were a test pilot named Zeubert, who flew the glider versions in the design stage, Granzow who was in charge of rocket development, and G Schaller – almost certainly a Nazi party member – who was 'co-ordinator' and responsible for keeping a watchful eye on progress at all times. The *Natter* was eventually designed as a single-seat rocket-powered intercep-tor. It was launched vertically from a ramp and flown almost straight up to meet the oncoming aircraft, under automatic control from ground-to-air radar supervised by an ordinary anti-aircraft superintendent. As soon as the enemy was sighted, the pilot took over and flew to within a few hundred yards. At this distance he fired a battery of rockets into the oncoming bomber and retired hastily.

The rockets carried were two sets of a dozen 73-mm anti-aircraft rockets of standard design, and after firing them the pilot allowed the plane to glide away to a speed of around 150 mph when he jettisoned the forward part of the craft and parachuted to the ground.

Officially known as the Bachem 8-349 A1, *Natter* was 4,800 lb in weight when fully laden, about 1,400 lb of that being fuel. The wing area was 46.0 square feet, and that of the tailplane 27 square feet; it was 21 feet 3 inches in length. Four Schmidding rocket boosters sent it into the air, each of

1 Höhenflosse 2 Tragflügel 3 Bug 4 Bugkappe 5 Vordere Haube
6 Panzerscheibe 7 Bugspant 8 Stringer 9 Raketenausstoßrohr
10 Obere Seitenflosse 11 Untere Seitenflosse 12 Seitenruder
13 Höhenruder 14 Rumpfheck 15 Deckel für Fallschirmkasten
16 Hintere Trennstelle 17 Rumpfmittelteil 18 Handlochdeckel
19 Belüftung für C-Stoffbehälter 20 Mittlere Haube 21 Panzerplatte
22 Fußsteuer 23 Salvengeschoß 24 Geschoß 25 Sprengbolzen

26 Rumpfspant 27 Hauptholm 28 Hilfsholm 29 Stringer 30 Höhenruder-
stoßstange 31 Antriebshebel 32 Fallschirmkasten 33 Bergungsschirm
34 Ausstoßvorrichtung 35 Sperrklinke 36 Seilzug für Sperrklinke 37 Auslöse-
vorrichtung für Bergungsschirm 38 T-Stoffbehälter 39 C-Stoffbehälter 40 Ein-
füllstutzen für C-Stoffbehälter 41 Belüftungsleitung 42 Rücklauf für C-Stoffbehälter 43 Rückenlehne 44 Sitz 45 Schütze 46 Bauchgurt 47 Schultergurt
48 Steuerknüppel 49 Sitzfallschirm 50 Rücklauf für T-Stoffbehälter 51 Einfüll-
stutzen für T-Stoffbehälter 52 Zuganker 53 T-Stoffentnahmestutzen 54 C-Stoff-
entnahmestutzen 55 Entlüftung für T-Stoffbehälter 56 Triebwerk 57 Höhen- und
Quersteuerung 58 Seitensteuerung 59 Innere Beplankung 60 Not-Sauerstoff

**Captured German drawings of *Natter* show the simple construction.
It took only 1,000 man-hours to build**

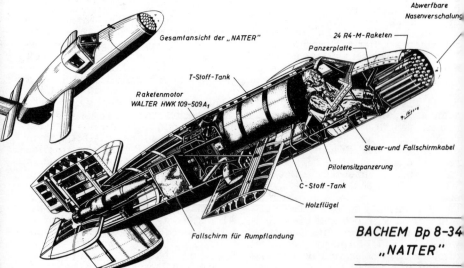

Gesamtansicht der „NATTER"

Abwerfbare
Nasenverschalung

24 R4-M-Raketen

Panzerplatte

T-Stoff-Tank

Raketenmotor
WALTER HWK 109–509A₁

Steuer- und Fallschirmkabel

Pilotensitzpanzerung

C-Stoff-Tank

Holzflügel

Fallschirm für Rumpflandung

BACHEM Bp 8-34
„NATTER"

Mit diesem Versuchsflugzeug füh
Oberleutnant LOTHAR SIEBER i
Februar 1945 den ersten senkre
ten bemannten Raketenstart de
Welt durch und fand dabei den To

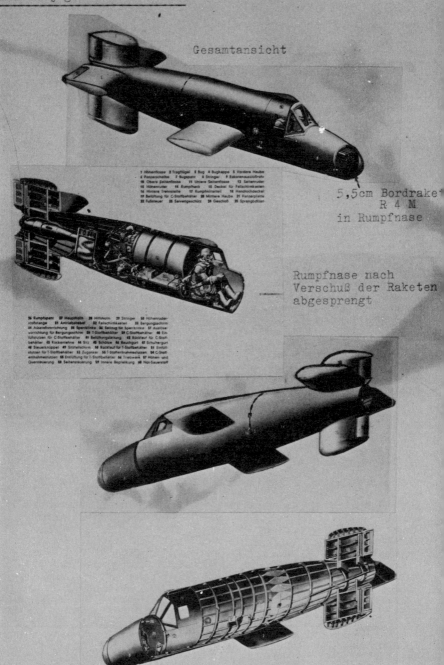

Gesamtansicht

1 Höhenflosse 2 Tragflügel 3 Bug 4 Bugkappe 5 Vordere Haube
6 Panzerscheibe 7 Bugsporn 8 Stringer 9 Raketenausstoßrohr
10 Obere Seitenflosse 11 Untere Seitenflosse 12 Seitenruder
13 Höhenruder 14 Rumpfheck 15 Deckel für Fallschirmkasten
16 Hintere Trennstelle 17 Rumpfmittelteil 18 Handlochdeckel
19 Befüllung für C-Stoffbehälter 20 Mittlere Haube 21 Panzerplatte
22 Fußsteuer 23 Sarvengeschütz 24 Geschoß 25 Sprengbüchse

5,5cm Bordrakete
R 4 M
in Rumpfnase

Rumpfnase nach
Verschuß der Raketen
abgesprengt

26 Rumpfspant 27 Hauptholm 28 Hilfsholm 29 Stringer 30 Höhenruder-
trofstange 31 Antriebshebel 32 Fallschirmkasten 33 Bergungzachirm
34 Auslaßvorrichtung 35 Sperrlinie 36 Sarvung für Sperrlinie 37 Auslöse-
vorrichtung für Bergungzachirm 38 T-Stoffbehälter 39 C-Stoffbehälter 40 Ein-
füllstutzen für C-Stoffbehälter 41 Befüllungsleitung 42 Rückstart für C-Stoffbehälter 43 Rückenlehne 44 Sitz 45 Schöße 46 Bauchgurt 47 Schultergurt
48 Steuerknüppel 49 Sitzfallschirm 50 Rückstart für T-Stoffbehälter 51 Einfüll-
stutzen für T-Stoffbehälter 52 Zuganlage 53 T-Stofferilrahmenstützen 54 C-Stoff-
entnahmestützen 55 Entlüftung für T-Stoffbehälter 56 Triebwerk 57 Höhen- und
Quersteuerung 58 Seitensteuerung 59 Innere Beplankung 60 Npz-Sauerstoff

26,400 lbs thrust and burning for ten seconds. The main power came from a Walter 509 A-2 rocket burning liquid fuels and producing a standard thrust of 3,740 lbs which could be throttled down to 330 lbs. The fuel consumption was stated to be of the order of 0.3527 ounces per 2.2 pounds of thrust per second at sea-level for 660 lbs of thrust.

Each of the armament rockets weighed 5.72 lbs and carried 14 ounces of powder as warhead. Later models were planned to have twice the armament capacity, though none of these were ever built. The maximum speed in level flight was about 500 mph and the range – after reaching $7\frac{1}{2}$ miles – was around 25 miles. The body of the aircraft was made from wood, nailed and glued unceremoniously together but designed to take a stress of 6g. The time taken to build each Natter was calculated at 1,000 man-hours.

Right from the start, when tests began, there were problems. In spite of an initial acceleration of 2g or more, the craft left the upright launcher at a speed of only 30-40 mph - far too slow to allow the wing flaps to exert any aerodynamic effect. Clearly (as in the V-2) some exhaust deflectors in the jet stream were needed. But how could they be included and still make the proposition economically sound? The answer was simple: steel vanes were included, each one hollow and filled with cooling water. There was no circulating pump or protective mechanism, indeed none was necessary. Admittedly after some seconds they would begin to heat up, then as the water boiled away they would eventually melt away in the rocket gases, but by then enough altitude had been reached for conventional controls to become effective.

And so an entirely new, and very viable, notion was introduced – disposable, temporary control surfaces. A dozen glide launches were made as tests of the basic airframe design; they took place at Braunschweig and proved very successful. Then came the question of manned tests under power. The sequence of operation was as follows:

After firing the rockets at the enemy, the pilot would dive and then glide down towards a safe landing area. He would then unharness him-self, lean forward, and pull a lever which detached the nose of the aircraft. Due to the aerodynamic lift over the upper curvature of its structure the whole front end would come away and fly off, leaving the pilot exposed to the slipstream. He then pulled a second lever which released a large drogue parachute on $\frac{1}{2}$-inch cables, attached to the rear of the Natter. The sudden deceleration threw the pilot forward and clear of the structure, which was left to land elsewhere by its own parachute and – it was hoped – be subsequently re-used.

However there was only one manned test flight under operational conditions, and it was a failure. After the Natter was perhaps 500 feet in the air the cockpit cover flew off with the headrest; it fell near the launching crew. The Natter continued up at a clearly uncontrolled angle of about 15 degrees to a height of about 5,000 feet, when it heeled over onto its back and in a flat spin hurtled down to the ground, smashing itself and its occupant to fragments. No further tests were carried out because the invading Allies were getting nearer. It eventually turned out that Botheder had a ski hut in Oberstaufen, ten miles from Isny, and named 'Einen Echalpe'. It was here that the design team had elected to rendezvous 'after the trouble is all over'. And so one of the Germans' enterprising schemes, begun too late to have any effect on the war or its course, came to a romantic and classically unpredictable end.

The rocket motor used for the Natter, the Walter engine, had been developed earlier by Hellmuth Walter in tests at the Neuhardenberg airfield, designed to develop a propulsion unit for a rocket-powered fighter, the Heinkel 176. Von Braun's group at Peenemünde carried matters further, with the development of a modification of the standard liquid oxygen/ alcohol propulsion unit used in the rockets themselves: indeed a von Braun motor is said to have been successfully used to power an He-112 fighter in 1937. Tests showed that the rocket motor could develop over 2,000 lbs of thrust for a period of half a minute.

However as the fortunes of war changed and as Poland was easily

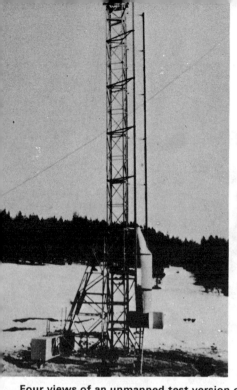

Four views of an unmanned test version of *Natter* mounted on the take-off tower. The only manned flight ended in disaster and death for the pilot

Below and above right: The first *Natter* mounted on a trolley before being transferred to the launching tower. *Below right:* A tricycle undercarriage was fitted experimentally but never used

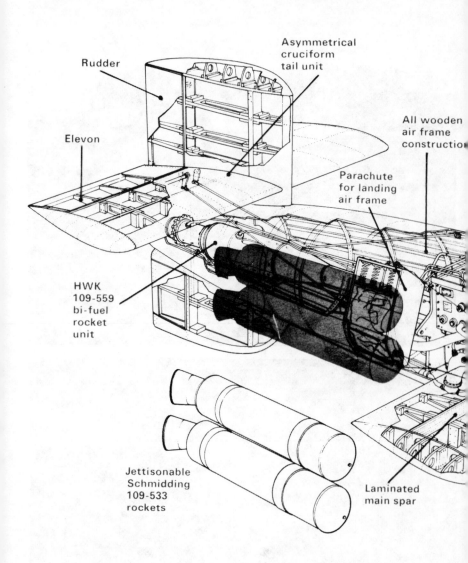

Rudder

Asymmetrical
cruciform
tail unit

Elevon

All wooden
air frame
constructio■

Parachute
for landing
air frame

HWK
109-559
bi-fuel
rocket
unit

Jettisonable
Schmidding
109-533
rockets

Laminated
main spar

Bachem BA-349A *Natter*
Crew: One. *Span:* 13 feet 1¼ inches. *Length:* 21 feet 3 inches. *Weight:* 4,800 lb.
Top speed: 560 mph. *Armament:* 24 55-mm rockets

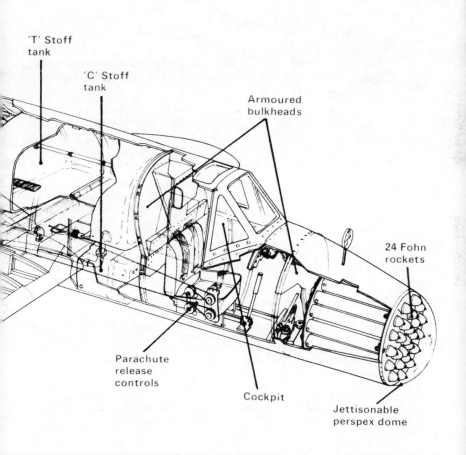

'T' Stoff tank

'C' Stoff tank

Armoured bulkheads

24 Fohn rockets

Parachute release controls

Cockpit

Jettisonable perspex dome

**The Messerschmitt 163A
rocket fighter**

overcome, the need for the project diminished and so – in the manner so typical of the government at that time – interest waned and the idea was cancelled. The He-176 died even before it was born.

But the concept remained, and a designer named Lippisch who had been working on rocket planes since the early 1930s decided to take it further. He had been working on delta aircraft for some years and by 1940 he had flown the Model 194, powered by a 660-lb thrust Walter engine, at speeds in excess of 300 mph. Then he designed a short, ungainly-looking aircraft, the Messerschmitt 163A, which was powered by the liquid-fuelled rocket. It was first tested in towed flights, but on 10th May 1941, Flight-Captain Dittmar flew it with its Walter rocket motor at speeds in excess of 600 mph. The world's first successful rocket fighter had arrived.

The design was interesting. The aircraft was built with a two-wheeled trolley undercarriage and tail wheel; as it flew up into the air at the end of its rapidly-accelerated runway boost

the undercarriage fell away, thus avoiding the aerodynamic drawback of exposed wheels and at the same time eliminating the need for heavy and cumbersome retracting apparatus to draw them up into the aeroplane's structure. The landing itself was done on skids, the Me-163A bumping roughly to a standstill along a grass landing strip.

It was not, by all accounts, a pleasant ride for the pilot. Films of the launching show the aircraft racing at a terrifying rate, oscillating somewhat and bumping wildly along, until it seemed to climb slowly above the ground whence it shed its wheels, leaving them to bounce along the field, leaping high into the air as they did so. Then, almost as though the projector speed has suddenly gone wrong, the film shows the aeroplane turning its nose to the sky and roaring up at a most impressive velocity. It was probably in the region of 10,000 feet per minute.

The aircraft itself is not an aesthetically pleasing design. Its surface finish was dull, rivet heads and screws exposed, and I confess that the first time I saw a preserved example I wondered how it could fly at all – the short, stubby wings seemed to lack coherence.

But fly it did, and remarkably well. The rocket fuel lasted for five or six minutes, during which time the aircraft reached its operational altitude and found the quarry. This was followed by a protracted glide back to the airfield – or, failing that, somewhere else flat and level – which took perhaps half an hour.

But designer Lippisch went further, and improved the design in many respects. His first design, back in 1939, had been for a tailless aircraft with wide, squared wings; now the Messerschmitt plant developed the Me-163 still further. The Mark 2 was built around a more powerful rocket motor fuelled by a mixture of hydrogen peroxide and 'C-stoff' – methyl alcohol, hydrazine hydrate, and water – which developed a thrust of considerably greater magnitude; rocket assisted take-off, to economise on fuel carrying capacity, was used for this, the Me-163B. Then the Junkers factory came into the picture, by building a Mark 3, which they called the Ju-263. It became known as the Me-263 after production reverted to the Messerschmitt factory.

However, though it would have been a formidable weapon of war, and certainly a great surprise for the Allies, it was never fully developed. Glide tests were carried out, but the end of the war precluded the Me-263 ever taking the air under its own power. Even so, its predecessors left behind them an impressive record for such an experimental aircraft: Lusar writes of Kapitan Olejnik who took off from Brandis airport near Leipzig late in 1944 and shot down several enemy bombers in the space of five minutes. From then onwards the aircraft was stationed at strategic points and was used to defend important plants against enemy bombing raids. The Leuna works were protected in this way with 'considerable success' it is said.

Altogether the Me-163B weighed just over 9,000 lbs; it was 19 feet 6 inches in length and could climb at over 10,000 feet per minute. It flew operationally at speeds near to 600 mph and had a 3,500-lb thrust engine which could be throttled down to 650 lbs of thrust. The model C would have been 4 feet longer and probably built for a crew of two.

However the basic success of the idea had been demonstrated, and from the drawing-board idea in the early stages of the war through to a successful, operational rocket-powered aircraft in use in 1944 is impressive as a record of war-time development. The subsequent development of the *Natter* is more difficult to see in perspective: it was obviously a rushed job, built hastily and with poor materials; one cannot easily decide whether it was in fact a triumph of expediency under trying circumstances or a panic move which was really too hasty to be worthwhile. The earlier experiences with rocket planes such as the Me-163B do show that the Germans realised there were practical benefits in the idea, and so perhaps the development of the *Natter* was not as thoroughly crackpot as some historians of the period have suggested. What is clear is that, had the war continued for longer, both the Germans and the Japanese would have had far more secret developments near to completion and – whatever the outcome – they would have proved to be more formidable opponents still. These secret rocket developments laid much of the foundations for post-war rocket aircraft, as it is.

Many of the secretly-developed aircraft that Germany used to surprise the Allies were derived from less unlikely sources. For example, the Heinkel He-111 had become a successful commercial transport aircraft before the war started, and its basic design was quickly adapted for military uses. It became the Luftwaffe's standard twin-engined bomber, and was produced eventually in literally scores of different types. Typical of them were dimensions of 82 feet wingspan, a length of some 52.5 feet and a speed of 280 mph. In 1940 some 750 of these aircraft were being built; the number had doubled by 1942-3.

A most strange aeroplane development took place in turn from this modified 111; this was a heavy-duty aircraft for towing heavy gliders. It was designed as a pair of He-111s literally stuck side by side, and sharing, therefore, a common wing on which a fifth engine was mounted.

Me-163
Crew: One. *Span:* 30 feet 7 inches. *Length:* 18 feet 8 inches. *Weight:* 9,500 lb.
Top speed: 596 mph. *Armament:* Two 30-mm cannon and 24 55-mm rockets

The entire wing span was nearly 125 feet, and the pilot sat in the left-hand cockpit controlling the plane through identical linked control systems.

Towing a glider of over 35 tons, the aircraft was recorded as having reached an altitude of over 30,000 feet in 1942. However the large amounts which would have been strategically necessary were never produced and, though the records are fragmentary now, there is no evidence that the aircraft proved to have been successful as was originally envisaged. Certainly the aerodynamic and handling properties of such a bastard giant would have been strange, to say the least!

There were many other aircraft which flooded from the Heinkel top-secret design workshops; after the He-111 came the He-115, a seaplane which was in some ways comparable with the concept of the Sunderland flying-boat – but this German idea came to naught in practice.

The He-116 was especially developed for long-range cargo transport; the He-117 and 118 were tactical developments which hardly got off the ground and the faster He-119 – capable of 375 mph and driven by two DB-603 engines – was, likewise, not accepted for mass production. However some of the design concepts were taken up by Japan for possible use in her own war effort.

But as the later war years made Germany realise that the Allies did stand a chance of victory, after all, she began to reconsider these dusty plans on the Heinkel shelves. The need for better, faster, more efficient bombers was a matter of great urgency for Germany in 1944, and a four-engined version of the range of designs developed from the He-111 was soon in the early stages of production. It had a bomb load totalling 17,500 lbs and was christened the *Greif* – the He-177.

But this new development failed. It was kept a complete secret during the planning stages, and was a subject of the highest security even afterwards, but as soon as it was near to production flying the news was spread throughout the German flying services

in order to boost morale. Germany has a world-beating top secret bomber, it was said; the enemy will at last meet its match. But this was a basic error of tact and psychology. Everyone's eyes were fixed on this new hope for the Reich; and the whole of the German political machine was stunned when the first of the aircraft mysteriously caught fire and crashed in flames. Later a second of the H-177s exploded in the air, and tests soon revealed the answer. The aircraft was overpowered, and the heat of its engines in operation was boiling the fuel in the adjacent wing tanks with obvious results.

Tests at the Heinkel works were carried out on the ground, and modifications were attempted. But of course an engine running on the ground is always in an unrealistic environment, for it is intended to have its cooling carried out predominantly by the slipstream of the aeroplane at speed. However, changes were introduced.

But the fault lay in the fact that the bomber – like any aircraft of its type – was far slower than other aircraft, such as fighters, could be; the slipstream was not enough to cool the engines in flight and they continued to heat up irrespective of added precautions taken in the design and insulation of their mountings and, although the modifications lessened the risks of an early failure, they were always liable to explode suddenly. Within months the model had been discredited in the eyes of the German hierarchy; and with it went the reputations of the officials who had sanctioned the production contract, and indeed the standing of the Heinkel firm itself. Later on a change of engine produced more satisfactory results, although there were difficulties then with the armament of the aircraft. This (probably due to over-enthusiasm by the designers, trying to compensate for their earlier blunders) was too much for the structure of the aircraft to stand and eventually smaller weapons had to be installed.

The Heinkel 111 bomber in action over England. This standard medium bomber of the Luftwaffe gave birth to several secret variants

So the He-177 staggered from one misfortune to another and, though England was bombed by the aircraft in 1944, the bulk of the 1,000-odd aircraft built went to the scrap heap. It was an ignominious end for a design project which was ill-conceived and rushed in so many ways, but which was in essence a logical development. Once again, German faddishness turned a potentially lethal secret weapon into a heap of scrap metal.

But still the ideas were there, and eventually they were bound to come to some sort of fruition. However, when this did occur, it was not in Germany at all.

The potential of the He-177 Mk IV was clearly great – if only it would not set itself alight. And so, although the production bomber itself was scrapped, many of the findings that resulted were put into designs for the He-274, which was, for diplomatic reasons, to be manufactured in Paris, at the Farman factory at 40 Avenue J J Rousseau, Sureance. This had been taken over by the occupation troops and converted to German military production. No-one there was taken completely into the Germans' confidence: when staff were sent from Paris to Germany for training or consultation they were whisked into the German offices, and – as soon as business was done – they were hustled rapidly back to Paris again. None of them ever saw factories in production, or were given any opportunity to find out about the overall German effort; clearly, even when the French were expected to help in developments, the Germans were so intent on absolute secrecy as to tell as few people as possible.

The new aircraft was reported to have a wing span of slightly over 130 feet and a length of half of this; the weight of the aircraft all-up was nearly 32 tons; range was to have been 1,875 miles with a payload of 2½ tons.

The Farman factory used a novel technique – the exhaust drive super-charger – to boost engine performance. The device took the form of a fir-tree-shaped turbine driven by the exhaust gases of the engine which in turn propelled an induction fan, driving the fuel mixture in under pressure and greatly increasing the engine's effi-

Crew members stand before the Heinkel 177 *Greif,* the Luftwaffe's ill-fated but potentially outstanding heavy bomber. It tended to burst into flames when heat from the engines exploded fuel in the wing tanks

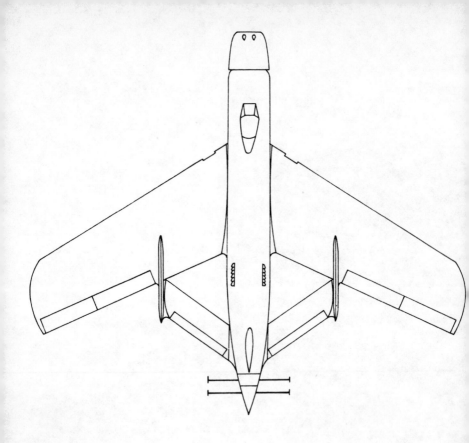

Focke-Wulf 03 10,251

Focke-Wulf 1,000 X 1,000 X 1,000

88

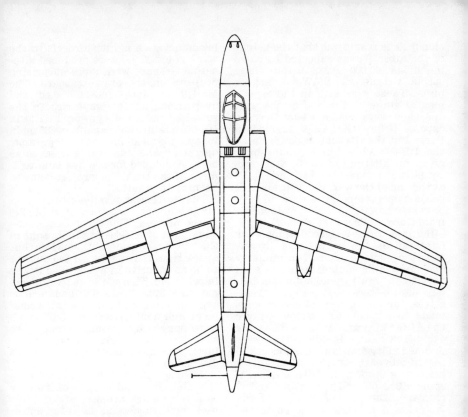

Focke-Wulf 03 10,025

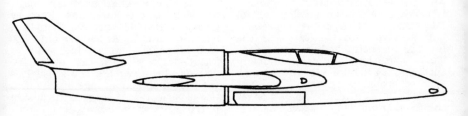

Messerschmitt P-1110

ciency. The result was that the height of the aircraft was expected to be as much as 40,000 feet, using these specially-designed BMW-801 engine units. These were still in the experimental stage at the end of the war, and they were destroyed by German experts before the Allies arrived in Paris. But the alternative power unit – the DB-603-A2 – which would have given an altitude of 20,000 feet, was actually being installed in the He-274 at the end of the war. Had this aircraft become operational, it too would have proved to be a formidable development from Germany's secret stable.

Heinkel aircraft also developed as fighters. In the early days it was reported that the German air ministry discouraged developments since such very fast aircraft would not be needed; but as matters rationalised themselves, further concerted efforts were made to outpace and outmanoeuvre the Allied aircraft. As a result, in the early 1940s, the He-219 series were planned. These began with successful test flights of the Mark 1: an all-metal high-wing monoplane with a speed of some 400 mph and a tricycle undercarriage. The Mark 2 was larger and faster; and a Mark 3 was envisaged too, which would have been powered with high-rating engines and thus have an improved performance – but the German ministry was dubious about their potentialities and so production was slow and mass-production lines were never established. Had this come to fruition it would have been a worthy opponent in the air, but, with German ideas running amok, it hardly left the planning stage.

But Heinkels made their mark, as we shall see a little later, in another important field – the development of jet engines of a most successful nature. Even so, the taint of the failures in their big bombers soured the reputation of the company throughout the later years of the war.

It is probably true to say that Messerschmitt aircraft were amongst the most prominent features of German activity in the war: they were far from being 'secret' in any sense. But here too there were security developments taking place which were maintained as closely guarded secrets. Though the Me-109 became

legendary as a fighter aircraft in the war – some 33,500 were produced altogether – there were more surprising projects on the drawing board. The Me-110 (a slow, underpowered and over-armed fighter) gave rise to the so-called 'Adolfine'. Powered by twin DB-610 engines and capable of 385 mph, it was planned for very long range operation and was even envisaged as being developed for use in a bombardment of the USA. The project came to nothing in practice.

The Me-110 was developed in another direction, too, becoming the Me-210 in the process. This incorporated the novel idea of self-sealing fuel tanks in the wings, a considerable development at the time although it turned out not to be as effective in practice as it was in principle. There followed a range of models – 310, 410 – which had various experimental refinements of armament and control systems but which were bugged by teething troubles to such an extent that – once again – further research and development was discontinued.

However one version of the Me-410 – the 'Hornet' – did prove satisfactory as a reconnaissance aircraft and was used with success as an interceptor. Messerschmitts also made important developments in the field of jet propulsion, a subject we will examine later in the book. These important research programmes, which gave Germany a world lead in turbine-powered aircraft, were completely unknown at the time in the West and, by any standards, must rank amongst the most significant of all the secret wartime developments.

Right at the other end of the scale was the super-small aeroplane built by the Arado Aeroplane Factory (based in Brandenburg and Potsdam, near Berlin). The Ar-231 was only 26.6 feet long and had a wing span of 33.5 feet; it weighed just over 2,200 pounds all-up. With a top speed less than 112 mph and a range of 280 miles it was perfect (in theory) for its design purpose – reconnaissance from a U-boat. The aircraft was fully collapsible and stowed into a 6-foot diameter canister on the stern deck of the U-boat; it was taken out and assembled in ten minutes or so (it was claimed) and then took off on sea-floats for an aerial survey. After its return it was quickly knocked down and stowed away again inside the container; then the U-boat, furnished with knowledge of enemy positions, dived away and in due course moved in to the attack on a completely unsuspecting ship. It was a ruse of classical proportions, but there was a snag. If the waves were more than a couple of feet in height it was difficult to get the aircraft on board again; and so this enterprising idea was disbanded and yet another of Germany's secrets ended on the scrap pile.

The submarine commanders knew, however, that in principle the idea was sound. The main limitation of the German U-boat campaign was the difficulty of reconnaissance: the distance of vision is severely restricted for an observer at sea-level and some means of gaining height – as in the abandoned aircraft project – was a good idea. Accordingly the effort was not altogether abandoned, but rather it was transferred to the idea of an observer suspended beneath a kite towed along on the surface by the U-boat. And not just a kite of conventional design, either; but a superbly designed and well thought-out rotary wing kite.

The project was put to the designers of the Focke-Achgelis Flugzeugbau (a division of the Weser Flugzeugwerke group) of Hoykenhamp, near Delmenhorst, who were already experienced in conventional helicopter production. The last conventional machine they had produced was the F-223, which was believed to have been made in 1942; but this was known more widely. It was their rotary kite which was the secret development, and which raised so many eyebrows amongst the Allies when the secret broke near the end of the war.

The Weser Flugzeugwerke were based at the Lloyd Building, Bremen, and they acted as government contractors for the deal: all development and manufacture, however, was carried out at Hoykenhamp. The plant superintendent – the man directly in charge of the efficient running of the operation – was Herr Fritz Kunner. A range of technologically advanced processes (such as magnesium welding) were in operation at the plant.

The kite was a fine piece of design. The entire device weighed no more than 180 lbs; the diameter of the rotor circle was 24 feet, the disc being loaded (with pilot) at about 75 lbs/ sq ft. Later models had an increased rotor length, taking the circle diameter up to 28 feet.

As the drawing shows, the construction was simple and effective. The main body consisted of a single steel tube on which all the other parts were welded. The small instrument panel at the front end carried a range of useful equipment, including an electric tachometer, compass fitting, and a telephone for communication with the commander of the U-boat. At the nose was the control array: rubber pedals and control stick to control altitude and inclination of the kite in flight. Small lateral projections supported skids which allowed the device to land on the U-boat's deck.

A second main member ran from the horizontal steel tube upwards, to carry the rotor attachments. The three rotor blades were places just forward of the centre of gravity of the kite as a whole. The linkages controlling the altitude of the rotor disc were cleverly designed so that they passed from the control stick to the rotor hub, through the steel pylon. Thus they were not distorted or damaged when stowed and allowed the kite to be completely folded away when not in use. The rotor hub itself was made of precision-welded steel tubing and contained an expanding brake in the centre which was used to stop the blades when stowage was necessary – sometimes under emergency conditions.

Underneath the rotor hub was a grooved wheel to take a rope which could be used to start the blades rotating. It was found in practice, however, that the pilot could generally reach up and give the rotor a hefty swing with his hand to start it. The starting-wheel was needed only when there was very little headwind.

The rotor blades were constructed according to the most up-to-date American and British methods at that time. They were made with 5-inch spacing between wooden spars, covered with very thin plywood (about 1/64 inch) along the leading edge. The

Focke-Achgelis FA-230 *Bachstelge,*
the rotary-wing kite which was towed
by U-boats to give their observers
greater height. Over 200 were built.
Weight: 180 lbs. *Rotor diameter:*
24 feet

whole of the rotor blades were covered with fabric glued to the members.

An ingenious device enabled the blades to be tilted in order to alter their angle of incidence into the slip-stream – hence their aerodynamic lift properties. Between the blades were high-tensile steel wires to prevent them from moving far out of spatial distribution and from the projections above the hub, similar wires ran out to support the weight of the blades and so to prevent them from 'drooping' unnecessarily.

The machine had an 'emergency' escape procedure if the submarine needed to dive suddenly. The rotor was detached when the pilot pulled a small lever above his head; the entire rotor assembly flew away, the parachute was pulled out of its stowage (behind the pylon) and automatically opened at the same instant. The pilot then released his seat-belt and the entire fuselage fell away into the sea.

To launch the 'gyro-kite' the U-boat would surface and head into the wind. The machine was quickly unpacked and fitted together with spring-loaded lugs and clips which easily slotted together by hand. The wing and tail surfaces were held in place by only two locations each, both quickly slotted into position; the collapsible seat and the telephone attachment were added and the towing cable (carrying the telephone lead) was clipped on. The machine was launched with a slight nose-up attitude and – either by using the rope on the starting-wheel or, more usually, by turning manually – the rotors were started. The minimum airspeed for a safe launch was about 20 mph when the blades·would begin to rotate at about 200 rpm, lifting the device off the deck. Gradually the cable was winched out as the 'gyro-kite' gained altitude, the pilot meanwhile sending back reports by telephone. It was intended that usually the device would be gradually winched down again after the flight was over; the blades would be halted by means of the hub brake, and the machine quickly stowed away. However on several occasions the emergency procedure was used instead.

This enterprising machine was nick-named the sandpiper – *Bachstelge* –

officially the FA-230. Altogether about 200 were constructed and they were used with great success by the U-boat crews who quickly learned how to pilot the simple gyro with its unsophisticated control system. The device was kept a complete secret from the Allies until the beginning of 1945, when one was sighted and reported in the British press. It is interesting to read in the official report on the factory when the Allies reached it later in 1945 that 'one of the rotors had a bullet hole through it ...' Clearly the Allies soon learned how to deal with Germany's U-boat kite.

Other more orthodox aircraft remained a secret to the Allies. Amongst them was the Focke-Wulf Falcon – the FW-187 – which was planned as a low-wing monoplane of all metal construction and armed with half-a-dozen machine guns. Powered by two Jumo engines it would have had an operational top speed of 330 mph – which made it nearly as fast as the Me-109. But it was cancelled by the Air Ministry whilst still in the development stage.

The TA-154 was another product of the Focke-Wulf stable. Designed as a fighter originally, it was diverted to unmanned control for use as a guided missile. Some were even tested using the Schmidt duct (exactly the same as used on the V-1 flying bombs) as an inexpensive means of propulsion; they were launched with a 'mother' aircraft nearby and so steered to the target. However there were formidable co-ordination difficulties in getting them all into the air safely – and at the right time – and eventually this idea too fell into disuse.

Then there was the FW-03 10.025: a grotesque design for a slender, tapering high-altitude fighter. Armed with two 30-mm and two 20-mm cannons, it was to have been powered by a 4,000 hp Argus engine amidships driving two contra-rotating propellors at the rear of the fuselage. The gracefully swept wings, spanning nearly 54 feet, made this a most sophisticated piece of design and echoes of its basic principles are still to be found in modern aircraft configurations. The FW-03 10251 was also pushed along by rear-mounted propellors; it had swept wings and forward-pointing tailplanes making it rather like a parallelogram from above. It was still in the design stage at the end of the war, as was the strange *Triebflügel* – an upright, vertical take-off coleopter. Its rounded, stout fuselage was surrounded about its midsection by three long arms, and at the end of each was a small jet propulsion unit. As they were switched on, the rotating arms – acting as blades of a helicopter – lifted the device from the ground and carried it swiftly aloft. The notion of VTO (vertical take-off) was the subject of a patent applied for on 10th September 1938 by the German engineer Otto Munch; it was another learned aerodynamicist, Professor Tank, who took the coleopter idea into some design detail at the Focke-Wulf establishment – but here too there was official apathy although, at the time, it seemed quite reasonable to expect that a velocity greater than that of sound was quite feasible for the aircraft. In fact no successful coleopter has ever been built.

Focke-Wulf also produced designs for the so-called 1,000x1,000x1,000 project, intended to bomb English cities. It was named thus since it was meant to carry 1,000 kilograms of bombs at 1,000 km/hour for 1,000 kilometers: equivalent to 2,200 lbs of bombs for 625 miles at 625 mph. With a wing span of 41 feet and a length of 47 feet, the near-delta aircraft clearly showed promise as a design entity – but would it have been successful in practice? As far as one can tell from the rough drawings remaining, it would have been heavy and perhaps unstable. In any event there was no opportunity for the Germans to find out in time.

Finally, from the Focke-Wulf designers, there was the projected FW-03 10225; a long-range bomber designed to attack the USA. It was intended to have a range of over 5,000 miles and a bomb load of 6,600 lbs. It had a large, bulky central body section with two fuselage booms running back from each of the two engines to support the tailplane. In this way the aerodynamic qualities were believed to be potentially improved, and the rear-turret gunner had a completely unimpeded view of the surrounding air-space. It was planned to fly the aircraft at altitudes near 30,000 ft at over 350 mph;

and, with armament of 8 or 9 cannons and 4 machine-guns, it was – in theory – a masterly development.

Other long-range projects centred on the novel idea of 'Mistel' mounting (meaning, literally, mistletoe – the analogy being that of a parasite on its host). This was the carrying of a small aircraft on a larger motherplane which would release it near to the target before returning to base: in this manner, it was felt, the attacking aircraft arrived on the scene of the target with a full fuel tank for the return journey. For the first of these trials (which earned nicknames such as 'Huckepack' (pick-a-back) or 'Father-and-son') an FW-190 was mounted onto a Ju-88 Mk 4. The latter carried a large hollow-charge bomb, capable of inflicting severe damage on detonation and ideal for the attack of standing installations. The pilot of the small Focke-Wulf had control of the two aircraft, the parent having had its cockpit completely gutted and frequently filled with explosives. During take off and flight the FW-190's propellor was feathered and only the Ju-88 provided the motive power. On reaching the target the device was very carefully aimed at the installation to be attacked, and the fighter then released the bomber and banked away. The Ju-88, meanwhile, plunged onwards towards the target, whilst the FW-190, its engine now switched on for the first time, headed smartly back to base. By the end of the war there were over 200 Mistel in existence; they were mainly used in vain attempts to halt the Allied advance.

But the most ambitious of the 'Mistel' configurations was never put into production at all. It was the projected Daimler-Benz 'A'; a giant low-wing bomber with a tailplane halfway up the vertical tail fin, and with widely spaced wheels lifting it some 20 feet into the air. Beneath the body and between these wheel mountings was to be slung a smaller fighter (with a V-shaped fin, to nestle it closely beneath the body of the parent aircraft) with a normal bomb load of 1,000 lbs. The aircraft flew together to the target under control from the cockpit of the larger, and the smaller was released near the target area for the final bombing attack. In this way a refreshed pilot with a full fuel load was delivered to the target, ready for battle and able to return to base after the attack was ended.

The V-shaped tailplane was a feature of the Messerschmitt P.1110 projected for 1944. A similar effect was produced in the Blohm und Voss P-208, a strange flying-wing machine with anhedral tips. But perhaps most prophetic of all the secret flying machines were the Lippisch series of delta-wing aircraft. It was Alexander Lippisch, the firm's founder, who first developed the idea of *Fliegende Dreiecke* – flying triangles – at Darmstadt. He built wooden fabric-covered gliders to test the

The Messerschmitt 262 jet fighter on the airport at Dübendorf

principle, the first being the small *Lilliput 65* which more than justified further research. There followed a series of tailless aircraft which were powered with various conventional and rocket motors. Lippisch knew a number of the rocket-power pioneers and apparently had assistance (and test flights) from several of them. Throughout the 1930s there were experiments and occasional disasters, but the basic soundness of the idea had been demonstrated.

After the war began, Lippisch continued to work by himself, aiming at the perfection of a design rather than the production of specifically hostile machinery. The first of his final series was the DM-1, an almost perfectly triangular aeroplane with a tapering

vertical tailplane which ran in virtually a straight line from the end of the
aircraft down to the blunt nose. It
weighed less than 1,100 pounds and was
less than 20 feet in length. In 1945 it
was given glide tests. It was towed to a
height of 12,000 feet or more and then
released to dive away at high speed. Its
stability and handling qualities
proved to surpass the most optimistic
of hopes and designs for a DM-2 were
on the drawing board when the war
ended. This would have been capable
of near-sonic velocities, and eventually it was planned to take it through
the sound barrier in level flight. But
it was this simple wooden glider, the
top-secret Lippisch DM-1, which pointed the way to much post-war development in aviation.

There was at the same time a powered rocket-aircraft, the DFS-228 Mk 1,
a high-altitude reconnaissance aircraft capable of reaching 60,000 ft (it
was hoped); powered with the Walter
rocket engine, it was towed to perhaps
25,000 ft and then after ignition it was
released and left to travel up to its
maximum height. The test flights
showed that the device was capable of
560 mph in level flight. After its reconnaissance project was completed, the
aircraft would have glided back to
base, its fuel supply exhausted. The
dozen or so models in existence as the
war drew to its close were destroyed by
the Germans, so it remained a truly
secret development; even so some
drawings and remnants gave a clear
picture of the aim of the project.

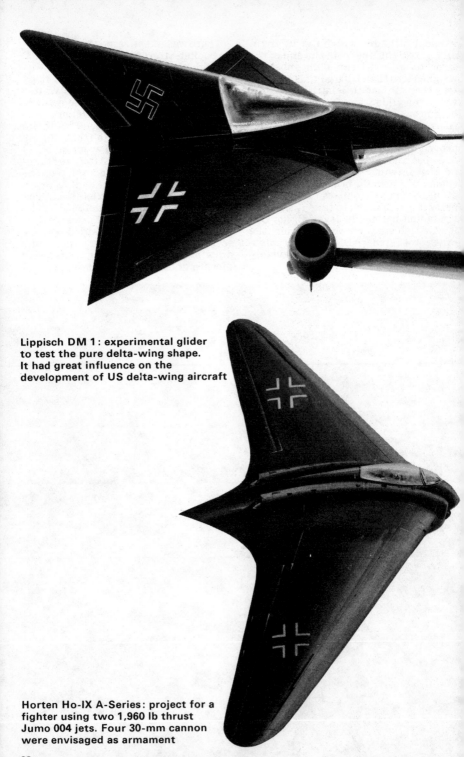

Lippisch DM 1: experimental glider to test the pure delta-wing shape. It had great influence on the development of US delta-wing aircraft

Horten Ho-IX A-Series: project for a fighter using two 1,960 lb thrust Jumo 004 jets. Four 30-mm cannon were envisaged as armament

Focke-Wulf *Triebflugel:* project for a coleopter fighter based on the vertical take-off idea

Without any question, however, the most bizarre of the successful secret projects were the bat-like flying wing aircraft designed by the Horten firm. In conditions of utmost security, the first of the test gliders had been built in Bonn by 1932. It had a wing span of 41 feet, a flying weight of only 440 lbs and as it glided in still air it lost altitude at only 2 feet 9 inches per second. It was brought out of its security shroud for the Rhon glider championships held in 1932, but when it won first prize it began to attract headlines and was subsequently burned, along with wooden models, to protect the idea as far as possible. Later models were made under a permanent cloak of secrecy.

The second of these flying wing devices – the H-2 – was flown in 1934 in both glider and powered versions. With a wing span of 54 feet and a far greater flying weight (830 lbs) it was none the less capable of a more maintained glide path, unpowered flights showing the rate of sinking to be 2 feet 7 inches per second – better than the Mark 1, in spite of the greater weight. The H-3 was built in Berlin, at Tempelhof airport, just before the beginning of the war. It was a little larger than the Mark 2, and lighter too. Metal skinning was used to cover the wing in parts as the plywood was proving too cumbersome and stresses were too great to allow fabric coverings. It sank during the free glide path at only 2 feet 2 inches per second. But this was improved yet again with the Mark 4, known as the RLM-251 which was constructed at Königsberg – Neumas in 1941. It sank through the air at only 1 foot 9 inches per second, in spite of a flying weight of 750 lbs. In the search for lightness, metal skinning was being used more and more, and subsequently a glider was built at Hersfeld which used light-weight plastic coverings for the wing surface.

The plastic was known as 'Tronal', and was especially produced for the project in thin sheets by Dynamit A G – Troisdorf. But the wings were too light and the stalling characteristics were unsatisfactory. The plane went into a fast spin during test flights and crashed to the ground in pieces, killing the pilot outright. He had apparently been pinned to his seat, unable to parachute to safety, by the centrifugal force of the wildly gyrating device as it plunged downwards out of control.

The Horten Mk 5 reached finality even sooner; it was projected and built at Osthein before 1939. It too had all-plastic wing coverings and showed considerable promise. But the pilot lost his nerve during the first flight and the flying wing crashed in pieces once again. The Mark 6 followed from the Mark 4; it had thin wings and a sharp leading edge but the wings drooped so much on the ground that the entire project was soon abandoned.

The H-7 was powered with two Argus Aslo-C engines each rated at 240 hp. It was built late in the war at Mindem and was test flown on the outskirts of Berlin at Oranienburg, where the flat sandy landscape was ideal for the purpose. There was an attempt to use a drag bar protruding from the wing tip as a directional control; acting as an aerodynamic 'spoiler', it proved, in flight tests, to be unreliable and unsatisfactory. The H-X version of the H-3 had moveable wing tips as a different approach to the problem, but this was not very much better than its predecessor.

But the end-product of all this research and development was the Horten H-IX-V2 fighter, powered by jet engines. It was originally designed and built by the two Horten brothers – Major Walter Horten and Oberleutnant Reimar Horten of the Luftwaffe – at Goingen. It was built for production purposes at the Luftwaffe's Sonderkommando-9 at Gottingen towards the very end of the war. Based on the H-5 (see above) the first of these designs was the H-IX-V1, an underpowered jet version; this was superceded by the H-IX-V2, essentially the same but with an alternative jet propulsion system of greater power. There was also an H-IX-V3 which was being constructed at Gothaer-Waggon Fabrik, and an H-IX-V4 which would have been a two-seater version with a larger and more pointed nose. The V1 and V2 versions both reached test flight capacity at Oranienburg; all detailed models and drawings of the later models were destroyed by burning before the Allied powers were able to see them. Thus in many ways they

The Dornier 335 fighter powered by two tandem-mounted engines

are destined to remain secret weapons in every sense.

The characteristics of the V2 version, as far as can be ascertained from the material still remaining, were impressive. The aircraft had a wing span of 53 feet 6 inches, a wing area of 452 square feet, and a weight at takeoff of 18,000 lbs. It was powered by two BMW jet engines and carried five fuel tanks in each wing. The aircraft carried four 37 mm cannons and two 2,200 lb bombs. The wheels were retracted inboard (in the V1 model the retraction was outboard) and there was a single nose wheel (two in the V1) which trailed rather like a furniture castor but which was not steerable. There were brakes on the main wheels and a spring operated ejector seat.

The wing loading must have been in the order of 40 lbs per square foot, and the maximum speed at 20,000 feet – with a full load – was quoted as being 720 mph. The endurance was in excess of four hours, and the landing speed only 90 mph. With a light loading, takeoff runs of 1,600 feet were obtained at Oranienburg, and (though no fully-laden tests were ever carried out) it was calculated that a run of 3,000 feet would be needed all up. The control system was ingenious; by moving the stick, the flaps would move in a co-ordinated manner to change both altitude and direction automatically; wing spoilers were controlled by foot pedal controls.

The entire machine was built of wood, except for the centre of each wing (which was made from welded steel tubing) and the thin wing tips (which were made of light alloy sheeting). The whole aircraft was lacquered with a smooth varnish finish to give aerodynamic coherence; it was claimed that wooden construction was utilised not simply because of a shortage of raw materials but because of its ease of manipulation. Wood technologies were at that time very much better evolved than those used in plastic or light-alloy production work. The initial flight tests were encouraging and it seems as though the Horten might have become a significant threat to Allied aircraft.

There was a far larger flying wing under construction at Luftwaffe Sonderkommando 9 at Gottingen as the war ended; it would have been ready for its first test flights in November 1945. It had a wing span of 157 feet and was designated the Horten H-VIII (the Mk 8) flying wing. It was intended to have a range of 4,500 miles, cruising at 200 mph at perhaps 10,000 ft; there was no pressurised cabin for the crew, which made altitudes greater than this impracticable, but power-assisted controls were being planned. This machine too was predominantly made from wood, and it was burned shortly before Allied experts arrived on the scene.

But these graceful, slender tapering aircraft – rather like Australian boomerangs in outline – were the most adventurous and technically advanced of their type at the time. They were good examples of the fervour for development which the perverted enthusiasms of Nazism provoked in the minds of the Germans at that time.

There was, as is so often the case, a less acceptable result to this rush towards newer, better, and more lethal secret weapons – the suicide plane. The Germans armed aircraft with explosives and put pilots in them knowing that their chances of survival were small; ostensibly they were meant to bale out before impact but they rarely did so. Indeed it has been reported by commentators in the past that the cockpit versions of the V-1 flying bomb were intended to be flown to the target by a pilot who would have been killed in the attempt. Several Allied reports spoke of German uniforms and pieces of flesh being found lodged in returning aircraft after severe collisions with such aircraft. But (compared with the Japanese, whose Kamikaze pilots set a terrible precedent as suicide pilots on the grand scale) the development was only of a limited nature: it was a panic move of a fearful, terrified power, fast losing control of a runaway situation.

Dornier aircraft became well-known during the war, but in this company too there were secret developments of equal inventiveness. The most remarkable of these was almost certainly the Do-335, powered with two

engines mounted tandem-fashion. Identical propellers were mounted at the front (in the conventional position) and also at the rear of the fuselage as a pusher-propellor. The hazard to the crew if they had to bale out was clearly great; so a small explosive charge was mounted just in front of the rear motor which could be detonated from the cockpit and blew away

The *Mistel* combination in which the pilot of the Fw-190 guided the unmanned bomb-laden Ju-88 to the target, released it, and returned to base

the entire tail unit in the event of an emergency. The aircraft had a modern style tricycle undercarriage and the tractor propellor at the front was reversible, so that it could be used as a brake on landing. This unusual and enterprising aircraft was destined originally to become a night bomber and fighter; it was eventually built in a range of experimental versions and won the nickname (probably on account of its strange outline) of 'anteater' – *Ameisenbar*.

So the secret German weapons of the air, from the grotesque flying wing to the tiny rocket-powered interceptors, were impressively varied in concept and design. They set many patterns which later designers were to follow.

The chemist enters the secret war

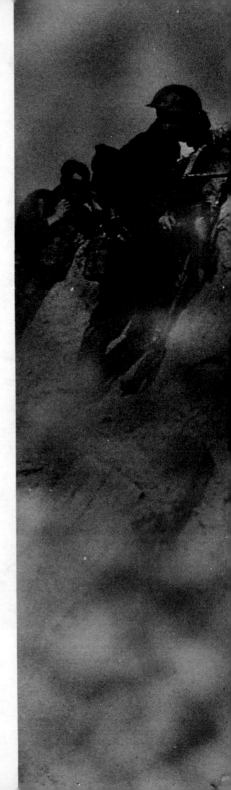

First World War gas attack

Chemical and biological warfare has always been feared above all. In its insidiousness and potential lethality it is the most devastating of the non-nuclear techniques of warmongering – yet it has an ancient history. In fact it was, in some ways, one of the oldest of the methods used to disable an entire army, in spite of the aura of modernity with which the subject seems to be surrounded.

In ancient history armies attempted to poison the water of a beleaguered city or a resting force of armed men by throwing diseased cattle carcases into the stream. Plague bacteria were encouraged to take over garrisons, and food poisoning was inflicted on victuals destined for consumption by the enemy. But many of these attempts failed in practice, and no doubt a good measure of the examples on record have been conveniently embellished with the passage of the centuries.

The first chemical weapon was 'Greek fire', a highly combustible chlorate mixture used from the 7th century until late in the Middle Ages in European conflicts – surely the prophetic antecedent of napalm. Until close to the end of the last century, the American Indians used incendiary materials in their attacks on the white settlers. Smoke was used by King Charles of Sweden's men when crossing the River Dvina in 1701 – mounds of damp straw were burned to obscure the enemy's view of the proceedings.

Gas warfare also goes back further than one tends to imagine, since sulphur dioxide (SO_2) was produced by burning an impure mixture of brimstone and pitch (i.e. sulphur and tar) as early as the war between Athens and Sparta in 431-404 BC. During the siege of Sebastopol in 1855, the English Lord Dundoland proposed that SO_2 should be produced by burning flowers of sulphur up-wind of the enemy, but the British government turned the idea down. It was, they resolved, inhumane. Much the same happened in the American Civil War when, in 1862, it was proposed that chlorine-filled shells should be used by the Union soldiers; this idea was also turned down.

At the Hague International Peace Conference in 1899, the subject was brought before the assembled delegates in the form of an official proposal. Britain had agreed to vote against the measure if there was unanimity amongst the rest of the countries represented – but the USA proved to be the stumbling block. Her delegates were not sure that the use of chemicals in warfare would necessarily be inhumane, as such; and so the United States voted against a total and permanent ban in principle. Britain's agreement with the proposition was conditional upon America's acceptance of it and therefore there was not the unanimity which had been hoped for. Britain would not support the motion; the USA was against it as a matter of principle. And so a world-wide concensus of opinion was not forthcoming in the unanimous sense that had been hoped for.

Just a few years later the world had a taste of the use of chemical materials when the First World War brought the use of poisonous gases out into the open. It was Germany that first set the precedent at that time, by using chlorine against the Russian troops on the Polish front in 1915. The Germans used mustard gas – 'Yellow Cross', as they called it – in 1917 and by the end of the war some estimates put the percentage of German shells filled with such materials as 50 per cent; the British were prepared to fill perhaps 20 per cent of their shells with the same material by 1918. Mustard gas was first used at Ypres: the French knew it as 'Ypérite' for some time afterwards.

So the scene was set for the Second World War. And Germany was actively engaged in discovering potent and lethal agents of war which would be used, if necessary, against the Allies. However, as we will discover, they were even more intent on defensive developments – Germany assumed (quite rightly) that the Allies had their own stockpile of chemical and biological weaponry, and she was determined to resist attacks from the other side by being well prepared for the event – an event which, as it happened, never materialised.

Perhaps it is useful to realise, when we speak of poison war gases, that not all of them were actually gases at all in the true sense. Many of them were solid material, but dispersed in a fine aerosol which acted virtually as a gas – they were, in effect, an ultra-fine dust. However they were not liable as were true gases to the adsorbent effects of gas-mask elements.

There are several groups of gases that are amenable to use in the theatre of war, and the Germans classified them according to four main subdivisions. They were:

'White Cross': gases such as bromine and its derivatives – such as chloraceophenone and bromine acetic ester – irritant materials but not of high lethality.

'Green Cross': these were the suffocating gases, like chlorine and phosgene, which attacked the lungs and caused death through pulmonary oedema.

'Blue Cross': this group consisted of the gases which block the respiratory system.

'Yellow Cross': these were the gases which had most dangerous surface effects – such as mustard gas and Lewisite.

And finally there were – always retained as top-secret weapons, subject to close security even today – the nerve gases which acted on the very transmission of nerve impulses. The first of these was discovered by the Germans and named Tabun; they were never used in operation. Let us examine these potent agents of destruction a little closer – in the first instance, how do they exert their malignant effects?

The choking gases of the 'Green, Cross' group are the most common and readily available of all these gases. They act purely as irritants and therefore enter the body at its weak points. In this case they act as they are inhaled, and they diffuse into the soft mucous membrane lining the respiratory tract; they also attack the delicate lung tissues. The physiological reaction of the cells in these regions is immediate; they are protected by watery secretions and these at once increase in order to attempt to wash the toxic material away from the tissues. This is normally a first-line defence of great efficacy, but the 'Green Cross' gases are such potent agents that the reflex secretion is

overstimulated and the outpouring of fluids is damaging to the lung: firstly because of the 'drowning' effect produced by the accumulations of the watery liquid and secondly due to cell damage brought about by the loss of water in this way. The irritant gases themselves subsequently cause chemical lesions to the cells and so the entire respiratory tract is fundamentally and often irreparably damaged.

Thus coughing begins, the lungs begin to become congested, the tract linings swell and expand into a watery mass, and death from a bizarre and lethal combination of suffocation and 'cellular drowning' occurs. Small exposures may allow the possibility of recovery, but anyone who has had a good lung-full of one of these gases is liable to suffer from some permanent disability.

Chlorine, which, as was mentioned earlier, first came to prominence in the First World War, is now an obsolete gas in warfare. It is a dense, greenish gas with a characteristic colour and odour and since it is markedly heavier than air it lies in hollows and in low-lying ground; alternatively, when propelled by a light breeze, it rolls along the ground in vast, eerie clouds of suffocating vapour; filling trenches, bunkers, crevices, or anywhere where men may be attempting to hide.

The chief area attacked by the gas is the capillary blood vessels in the lungs; then the lung tissue swells into a spongy wet mass and gas-exchange becomes virtually impossible. Death from suffocation results. Diphosgene is a similar material which is in the form of a clear, volatile liquid. Its effects may be delayed for perhaps some hours and it is fairly persistent in its action. Carbonyl chloride is another of the gases considered for use by the German scientists; it is more immediate in its effect and it passes off more rapidly. Like phosgene and diphosgene, which share a characteristic odour of freshly-cut grass, it was potentially available for use in the war.

The tear gases – 'White Cross' – produce a similar form of irritation, but here it is confined principally to the upper respiratory tract and the eyes. These gases are not persistent, and the effects pass off within a short while of the patient being brought into fresh, uncontaminated air. Tear-gases are generally used for civil disturbances, of course; but there are more potent compounds. One of these, Adamsite, acts initially as a classical tear gas but the effects are more persistent: within a few minutes this yellowish solid material produces coughing of a more violent nature, followed by a severe, blinding headache; violent spasms and pains in the chest produce severe difficulty in breathing and the nausea and vomiting begin. The gas persists in air for only about ten minutes, and is therefore classed as 'non-persistent'; but the effects last for hours or days and are far from pleasant.

It is the 'Yellow Cross' gases which produce the most disfigurement. Known to the Allies as 'blister gases', more correctly as vesicants, they attack any part of the body with which they come into contact, producing burns and blisters and deep, slow-healing ulcers of a most painful nature.

More systemic effects may later appear as the materials tend to interrupt the normal processes of cell division in the tissues and, in addition to producing wounds which cannot easily heal, it also interferes with the protection mechanisms of the body by delaying normal cell replacement and proliferation.

Distilled-mustard is a colourless liquid with an odour of garlic, although when prepared commercially its colour is generally slightly yellow. It is a most persistent agent, and after its vapourisation in an area it may persist for a matter of days. A related compound is methyl *bis*-amine hydrochloride, more generally known as nitrogen mustard, a dark-coloured liquid which may persist for most of a day after release, and produces severe blisters and burns on the exposed skin surface. The effects of these gases may take some little while to appear.

More rapid in its effect is Lewisite, an oily fluid which has the effects of nitrogen mustard and, in addition, the propensity to produce lung oedema and pneumonia which is usually chronically debilitating and often lethal. It has a musty odour which varies with the producer of the material, but which is said to be quite unmistakeable in practice. Indeed when gas has been used in the past it has always been the faint whiff of the substance approaching in the air which has been used, not only to warn of the imminence of the gas, but also to diagnose its nature. All too often, of course, the warning period proved to be far too short.

The 'Blue Cross' agents are the gases which block the oxygen carrying capacity of the blood. Arsine, the hydride of arsenic, is a gas with a smell of garlic – once again – which is highly dangerous. But most lethal of all are cyanogen chloride, cyanide, and carbon monoxide. All interfere with the up-take of oxygen by the haemoglobin of the blood stream and thus a form of 'physiological suffocation' takes place. The poison forms a molecular attachment at the site in the haemoglobin molecule where oxygen is normally carried. As a result, the life-giving gas can no longer reach the cells of the body and death is inevitable once this occurs. The soldier may gasp for air, and indeed inhale it; but his cells will not receive the oxygen they need for life and the victim soon succumbs.

The bonding of the gas with the haemoglobin molecule is, in the main, irreversible. Thus, unless the patient can be brought out of contact with the gas and into fresh air immediately, his capacity for life is steadily diminishing. Unlike the gases which cause suffocation – the 'Green Cross' gases as the Germans classified them – the patient is not an unhealthy pallid colour. It is the combination of the haemoglobin molecule which causes its bright red colour and, even though it has combined with a poisonous gas instead of oxygen, the blood is a bright, almost cherry red colour.

In the field of battle the gases would be of low persistence and they rapidly disperse – principally because of their low densities. There are differences between the action of arsine and the others in the group; arsine attacks the liver and the kidneys as well as the blood stream whereas the other gases tend to damage the central nervous system instead. Moreover, arsine is often able to produce its first effects long after contact – the longest delay being over ten days.

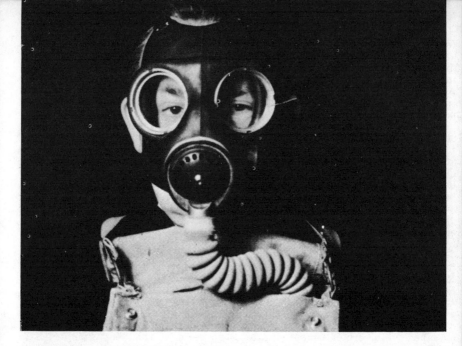

British (left) and French (right) gasmasks issued during the Second World War but never used in earnest

Finally there is a group of gases of unprecedented danger. They were never given a 'Cross' class name by the Germans for the simple reason that they were the object of total secrecy. For security reasons (these gases are of great strategic importance today) many of the details are still not cleared for publication and a range of more recent developments have carried their sophistication even further. The first to be discovered was the German compound which they christened Tabun. Chemically it was cyanodimethylaminethoxphophine oxide, and the Germans probably developed it about 1939-1940. They soon followed it up with Sarin, or fluoroisopropoxymethylphosphine oxide; and later developed Soman, which was fluoromethylpinacolyloxyphosphine oxide. The names are forbidding. So were the developments they described.

The three materials were colourless or pale liquids, sometimes brownish if impure. All acted within minutes or, if the concentration was high enough, immediately, and they ranged from being slightly persistent, in the case of Sarin, to being persistent as with Soman. This latter gas was the most

dangerous of them all; it was far more effective than the others and, if present in similar concentrations, its effects were far more violent and certain. How did these dangerous materials exert their terrible effects?

The whole of the body's functions depend on the transmission of nerve impulses along the various sensory and effector pathways of the body – i.e. through the communications network of the tissues. The impulses themselves pass through the action of a complicated series of chemical and electrical changes which in turn lead to the production of a compound at the nerve junctions. This compound is acetyl choline. The nerve is restored to its 'conductive' state by the removal of the acetyle choline by an enzyme which attacks it; the enzyme, cholinesterase, is therefore a vitally important key to the processes by which the body continues to transmit nervous impulses.

The nerve gases act by phosphorylating this vital enzyme – by adding a chemical group to the molecule which destroys it, or at least by preventing it from acting as it should. Thus the vital return mechanism is

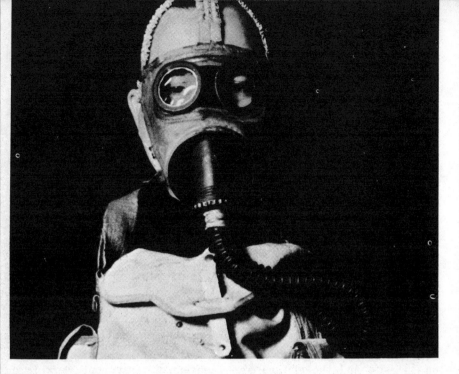

destroyed; the nerves are unable to continue to pass stimuli from one part of the body to the next, and the entire functions of the system become erratic and incoherent. On inhalation, nausea, vomiting, and diarrhoea occur; muscular twitching comes on rapidly, there are systemic changes in the blood picture, and convulsions soon ensue. Even the smallest traces cause the pupils to contract, making vision indistinct and incapacitating the soldier on the field of battle. Soman has the most terrible effects on soldiers: within seconds they would have been reduced to a state of convulsive collapse and death would be relatively certain by that time. It was only after the war that the effects of these materials were uncovered in any detail. The US Army Technical Manual TM3-215 lists the following effects of the nerve gases, in the order in which they occur:

Running nose,
A tightness of the chest which interferes with breathing,
Severe contraction of the pupils resulting in distorted vision,
Severe difficulty in breathing,
Salivation and heavy perspiration,
Nausea and vomiting,
Painful cramps, accompanied by involuntary urination and defecation,
Twitching of the muscles, jerky and unco-ordinated movements of the limbs, a staggering gait,
Collapse or coma,
Convulsions, total cessation of breathing, and death.

When a nerve gas is absorbed through the skin in effective quantities, the effects come on rapidly and death occurs within one or two minutes. With smaller doses coma and death may be delayed for over an hour, as the effects listed gradually occur over that period; but if the gas comes into contact with the membranes of the mouth or the eye then the sequence takes place rapidly and death occurs within ten minutes.

At the beginning of the war, phosgene was always regarded as the most dangerous of the war gases; the danger of death from Sarin is over thirty times as great. One-tenth of a milligram of this gas (equivalent to a particle the size of a large grain of sand) is enough to kill a child and three-quarters of a milligram is fatal to a fully-grown adult. Even so, it has been calculated that some 250 tons would have to be distributed over a city the size of Paris to cause lethal concentrations up to an altitude of 50 feet.

Experiments to develop these gases went on throughout the war and many of them took place at the concentration camps. Natzweiller-Struthof and Sachsenhausen were two of the first of these appalling establishments where such perverted experiments were conducted; later a euphemistically-named Practical Research Institute in Military Science was employed on experiments into the effects of poison war gases. Professor August Hirt, of the Department of Anatomy, University of Strasbourg, carried out the first trial, at Natzweiller. After a fortnight of 'acclimatisation' to the surroundings, concentration-camp prisoners were given doses of the gases on the forearm (the materials were of the mustard-gas variety). Within a day, large blistered burns had appeared on the site and the victims were photographed at all angles to provide a record for the authorities – who were asking for evidence of results for the expenditure.

After five days from the administration of the gas material, the first deaths occurred; the others followed later from delayed physiological effects and also from intercurrent infections which rapidly sprang up and spread through the weakened tissues. Post-mortem examinations showed that the 'internal organs had been rotted away', the structure of the lungs being 'similar to withered apples'. It is a dreadful record of distorted, inhuman torture – and all in the name of scientific research.

But at the end of it all the Germans were well armed with these war gases as a standby possibility: over 7,000 tons of Sarin alone had been stockpiled by the end of the war. And that is enough to completely kill the occupants of thirty or more cities the size of Paris.

On the biological front, many species of bacteria were investigated for their potential use as secret, hard-hitting weapons of war. Clostridium botulinum, for instance, can be cultured in large amounts and produces the most poisonous substance known to mankind. This is the toxin, botulin, which the bacteria produce as a side-effect, a waste product, of their norma-metabolism. Exact estimates of the lethality of the toxin vary, but a pound or two of the toxin would be able to wipe out an enemy entirely. The Germans knew of Clostridium botulinum and its potentialities, but they did not actually press ahead with the preparation of weapons based on the knowledge. However the possibility of cruising up the west coast of Britain, liberating a cloud of death which would drift across the island on the prevailing winds, was quite clear to the German government. In the first place, they had authorised research into the production of aerosol sprays which were capable of releasing the agents into the air in the form of an ultra-fine mist of minute droplets.

After the invasion of Belgium by German forces, it came to the attention of the government officers concerned, that in Brussels there was a university department working on aerosol production. The work was spearheaded by Professor L Dautrebande who lived in Chausée de la Grande Espinette in the Rhode St Genève district of the city.

The professor indicated that he had been working on aerosols for some time; indeed the Germans knew of his publications in that field [see bibliography]. He had coined the phrase 'aerosols vecteurs' and described the way in which an aerosol remains stable: it was, he stated, because of the ultra-small size of the particles which acquired an electrical charge and therefore (since 'like repels like') they would never separate out, since the particles would be quite unable to coalesce. The professor had carried out his work from the standpoint of one who wished to disperse therapeutic agents in the air of a room for the purposes of treatment – but to the Germans there was a more sinister application which seemed to be possible.

The device he had developed was a cylinder several inches tall; at the base of which was a high-pressure air inlet which entered the cylinder through two small apertures roughly 1 inch above the liquid layer in the bottom of the device. The air pressure swept up small particles into the aerosol form, together with a proportion of larger droplets which were uncharged and therefore unstable. Thus there was a percentage of 'mist' in the aerosol, and this had first to be eliminated. Professor Dautrebande had therefore included in his design a series of eight or ten perforated discs acting as droplet-traps or filters; in this way the majority of the larger particles were eliminated and ran back into the fluid, resulting in an aerosol of 97 per cent purity to be achieved.

At the department in Brussels, during the German occupation, a series of experiments was carried out in which various gases (principally of the 'Yellow Cross' group – but also nerve gases) were atomised in the device and fed into cages of rats. It was found to be possible to kill animals in this way purely from oedema of the lung tissues, without any trace of external damage at all. Clearly this was a most efficient way of releasing these lethal materials into the atmosphere. Had it been used operationally the consequences would have been incalculable.

This work had in fact begun even before the war, and the preliminary findings were certainly known about and read with interest by the Germans. The Germans also concentrated on means of protection against attack. Thus – under Dr Stampe – filters against gases such as cyanide were evolved; in this case pumice-stone was impregnated with copper-salt solution, then dried; and then re-impregnated with caustic alkali and dried again. This was shown in tests to be effective against the gas. For arsine, silver nitrate impregnated charcoal was used with efficacy. In other fields of interest, test units were developed to detect the smallest traces of poison gases. Thus a mixture of gold chloride with either chloramine-T or (in later models) sodium p-nitrophenyl-antidiacetate was found to be extremely sensitive to mustard gas.

All of which poses the most pertinent question of all – why was chemical and biological warfare avoided so particularly by the two sides of the 1939-45 conflict?

Almost certainly, I believe, it was because of the technological stalemate that existed. The Germans knew all that there was to know about chemical and biological warfare, but they were convinced that the British and American experts were as far advanced. Inevitably, they felt, retaliation would be swift and probably devastating; as a result there was no point in taking the risk. Certainly the Germans found that (in the high-density bombing raids which she pioneered so successfully, for instance) the Allies were well able to hit back hard. The appalling consequences of an all-out attack with war gases and bacterial poisons were too severe to be considered.

But here too there were interesting and, in some cases important developments that resulted, in particular the plant built at Seelze in 1935 for the production of the tear gas chloracetophenone (CN). The first installation was only a pilot plant but by the outbreak of war over 100 tons per month were being produced and stockpiled. The process (managed by a Dr Heineman in later years) was a masterpiece of well-planned chemical engineering, and purity of working air and of the plant's effluents was ensured by a number of ingenious developments. Of course, most of the enterprise was wasted, as the need for the gas never materialised. As a result the Seelze plant was closed down and put only on 'standby' in 1940; it was restarted for a while in 1941 but was wrecked by an explosion (probably the work of counter-Axis Intelligence agents) seven weeks later. Subsequently it was rebuilt, but never used.

But it was the fate of its 'sister-plant' at Leese, intended for emergency use, which is the most revealing. As it became apparent that it was never going to serve a useful purpose it was necessary to justify its existence in some other way.

And so this plant, built by J Reidel and E de Haen AG, the Hanover-based manufacturers of the illustrious Reich's tear-gas supplies for the war effort, was slightly modified. In the event it never produced a single gram of chloracetophenone, but was used to make vanilla flavouring instead – and saccharin tablets.

Spreading the net

Secrets are no use unless they are kept, of course. But (and perhaps this point of view is less widely emphasised) they have to be worth keeping in the first place. In war the secrets that are worthy of the title are not merely new developments, recently adapted weapons, or chance discoveries of military significance; they must represent a vast, orderly advance in many fields of endeavour – interdisciplinary endeavour, at that.

The German effort had this polyvalent quality; it embraced the whole field from metallurgy and chemical engineering to radio and television. And even in relatively humble applications there were important strides to be taken. The Germans were finding new plastics, they were making rapid progress in radar, as we shall see; but even in the field of basic engineering design there was much to be done – and much which, in the event, was achieved.

Before moving on to consider some of the more exotic of Germany's projected weapons, we may consider one of their secret devices which embodied exactly this breadth of vision in the development of a single item. The device in question was a torpedo motor: Germany had a need for high-efficiency, reliable torpedos – yet their power units were always a thorny problem. Junkers, by utilising a rotary valve concept developed by the Wankel company (which has, since the war, become world-famous for the design of an entire rotary petrol engine) overcame these formidable difficulties. The project itself was a demanding one, and needed all available expertise to solve the many problems involved. In the first place, for reasons of economy and power output related to size and weight, an internal combustion engine had to be the answer. But it was to run under water, and so could not use air; it had to reach full load and maximum output in the shortest time, and quite automatically; it had to be capable of driving the bulky torpedo at 40 knots or more with minimum weight – and because of size limitations the use of conventional valve gear seemed impossible. Crankshafts, bearings, transmission gear etc had to be strongly built and capable of withstanding the high stresses of rapid start from cold – say, from standing to full power within two seconds – and the accessibility of the engine ought to be suitably facilitated for last-minute overhauls. On the other hand, a short life of the components was clearly adequate for the job in hand.

The answer to these formidable problems was the Jumo KM 8 engine. Instead of air, it burned its own exhaust gases, topped up with oxygen, and petrol; and instead of conventional rockers and valves in the cylinder head it had flat plates with opposed apertures which opened the inlet and exhaust ports when the wheels rotated. The engine was a water-cooled V8 arrangement with a compression ratio of 6.6:1 and a total swept capacity of 4.34 litres (265 cubic inches). The complete unit weighed 450 pounds, and its maximum output on the bench at 4,360 rpm proved to be 425 hp. It was fitted with the Bosch twin magneto ZI 8, a pair of Bosch spark plugs in each cylinder – standard W 240 aircraft plugs – and a simple single-jet carburettor. A Graetz AG type ZD 53 double gear fuel pump with a maximum output of 77 gallons per hour was fitted. The crankcase was made of an advanced light alloy mostly aluminium but with silicon, manganese, magnesium, copper, zinc, and titanium included in small amounts.

The crankcase and cylinder blocks were an integral alloy casting, with liners screwed, and then pressed, into place; the whole assembly was held together with studs. As in the most modern V engines there was a main bearing between each of the big-end members. Water under pressure from a pump unit capable of 66 gallons per minute was used to cool the very hot surfaces produced by the explosive action of the oxygen on the fuel, and the disc valves were found to be capable of enduring the high stresses involved in spite of their simple design and construction. The engine was run on the bench – with air instead of oxygen and exhaust fumes – for over 50 hours without any trouble developing, so this was clearly a very much more successful device than. might have been anticipated.

It was a masterpice of applied design in engineering, and would certainly have been a most dangerous secret weapon had the torpedo ever been completed. But, as was so often the case in the Nazis' muddleheaded organisation, the development was neglected. At first an order for a hundred of them was placed, for delivery early in 1945; but then it was cancelled at the last minute. And so some eight years of development was ignored and thrown to waste by the government – one might add, greatly to the benefit of the Allies.

As the Germans' search for newer, more devastating weapons gained in intensity throughout the war, individual commercial concerns were encouraged to spread the net ever wider. The activities of the Draegerwerk company in Lübeck, for example, illustrate the variety: the group encompassed parachute oxygen apparatus, pressure suits, air-purification systems, submarines, silent motors, underwater escape gear, air-raid shelters gas masks, detecting apparatus for poison gas, and even an anti-dimming fluid to treat windshields – an impressive range of technological innovation.

One of the group's most significant, if small, fields of progress was the air purification apparatus they evolved for submarine use. It was designed to draw air from a midget submarine through a mixture known as 'Kalk-patrone' (soda-lime) and then, after the addition of extra oxygen from a cylinder of the gas, the mixture was returned through a rubber pipe to the occupants. If the 'pilot' of such a midget submarine wished to conserve oxygen, it was possible for him to plug a mask-hose on an adaptor on the outlet tube and then breathe in through the special injector apparatus and the soda-lime container, thus taking only the oxygen needed for his personal requirements. He breathed out through the mask, which contained a one-way valve letting the waste gases straight out into the compartment. Thus, though oxygen economies were effected, carbon dioxide accumulated in the atmosphere – although it was, naturally, removed by passage over the soda-lime from the air he was about to inhale. This was not a good prin-

The assembly line for production of
Seehund **two-man submarines**

Seehund two-man submarines mounted on trailers for the journey from the factory to the port

Seehund midget submarine:
Displacement: 15 tons. *Speed:* 7¾ knots.
Range (with external fuel tanks):
500 miles. *Crew:* Two: *Armament:*
Two 21-inch torpedoes

ciple, and in later models of this type it was found to be preferable to allow the exhaled gases to be passed through the soda-lime before returning to the compartment. In this manner the carbon dioxide levels were kept low.

Carbon dioxide, in small amounts, actually stimulates the respiratory reflexes of the tissues and thus the inhalation of the gas from cabin atmosphere will secondarily tend to increase oxygen consumption still more.

A development of this type may seem to be small but in the essentially practical nature of secret weapon development it turned out to be vital. For it enabled the Germans to press ahead with the production of a range of small – in some cases minute – miniature submarines and manned torpedos. And, using the expertise developed on the earlier projects, it was the Draegerwerk group which developed the most impressive range of them.

The first to roll off the assembly lines at the beginning of the war was the *Hecht* type, of which about fifty were produced altogether. Most were used for purely experimental work and trouble-shooting of control problems, although it seems certain that some were eventually used operationally. They were built as torpedos, but with a detachable warhead; they were powered by an electric motor – not of great efficiency – and, since the injector purifier described above had not then been developed, the 'pilot' had to wear his oxygen mask throughout and breathe oxygen. Next on the list was the 'Mother-and-child' project – the *Neger*. It had a petrol-driven engine and was also built on the torpedo principle, but in this case the manned unit remained intact: the torpedo proper was slung beneath it. Again the 'pilot' had to breathe oxygen, which proved to be an annoyance in practice; and so the *Neger* was modified and for the first time an injector system for the air supply was installed. The Mark 2 *Neger*, christened the *Molch*, was more successful in practice. Similar craft, known as the *Marder* series, were also constructed along the same general lines but powered by electric motors rather than the internal combustion engine of the *Neger/Molch* type.

The success of these small manned torpedos encouraged the designers to press ahead with something nearer to a miniature submarine in principle. The first design was the *Hai*, and it was a marked improvement.

The *Hai* never reached the production line, but it demonstrated the immensely practical nature of the miniature submarine concept. Flattened sideways, like a metallic sardine, it was powered by an internal combustion engine (a diesel was tried in some models, with commendable success) and could reach a speed as high as 22 knots under water. This speed could be maintained for up to two hours, although the shape made surface manoeuvres difficult and these limited speed to 7 knots at the most. It was able to stay down for up to sixty hours, and it was the injector apparatus which enabled the air to be kept pure and wholesome for such a protracted dive; without the earlier research, therefore, this first miniature submarine experiment would have been impracticable – or at least superfluous, since lengthy dives would clearly have been precluded.

This led inexorably to a more submarine-like design, the *Biber*. At first it was designed for a single 'pilot', but later a two-man version was developed at the Lübeck Flenderwerke plant. Two electrically driven torpedos were carried in recesses set into the keel beneath the paired bulges running along the hull. In these were the fuel tanks for the engine. But it was the *Seehund* which was the most successful of all the secret miniature submarines, and it went into production at the Neustadt Kleinverbande works and at Germania Werft, of Kiel.

In this field, too, there were plans afoot at the end of the war which time had prevented from coming to finality; Dr Walter, the designer of the peroxide rocket motors which were considered earlier, was designing a petrol/H_2O_2 turbine to make a jet-propelled miniature craft possible. It would have had a speed of perhaps 60 knots under water, in theory at least. That, in any language, is travelling. But would it have worked? In the opinion of several Allied experts who examined the plans and models at the war's end, it certainly would. It is only the change in the nature of 'personal warfare' which has prevented the widespread development of the idea further.

The Draegerwerk group, thus, were responsible for many varied programmes of research and were behind many of the advances responsible for the progress made in U-boat warfare. However there were many different workers investigating the secret war at sea, and as the need for the blockade strategem became clearer and clearer to the German government, hopes came to centre increasingly on the torpedo.

At the beginning of the war the Germans were equipped with a torpedo design which they felt to be adequate; it was fitted with a proximity fuze which responded to the magnetic field surrounding the iron hull of a ship. The torpedos were launched on course and travelled at a depth of a fathom or two to the target, exploding as they passed beneath it and breaking its back. In the theatre of naval warfare, however, these propensities were not fully realised since the fuze system was seen to be unreliable, and the depth-controlling fins were inadequately controlled, resulting in the torpedo tending to pass beneath the Allied vessel's hull at too great a depth for the torpedo to detonate.

The British scientists, too, soon discovered the secret of the magnetic sensors in the German torpedos and were able to fit electrical degaussing apparatus to sea-going vessels which removed the magnetic field and so invalidated the German torpedos altogether.

Many different designs were tried in the effort to make the German torpedos as efficient and reliable as possible, including acoustically-operated fuzes and automatic steering systems. Some of these enabled the submarine to fire the torpedo in a different direction from its eventual trajectory; others led it in an irregular path through a convoy knowing that one ship was likely to be eventually damaged or destroyed as a result.

The *Neger* 'mother-and-child' project, showing the upper hull containing pilot and engine, with the torpedo slung underneath

British officers examine a captured
Seehund miniature submarine

Above: Schnorkel mounted on a U-boat. This highly successful device enabled U-boats to recharge their batteries with considerably less danger of detection by radar.
Right: Schnorkel in use at sea

Many of these devices were not truly secret weapons, since they were sighted from ships and their path of attack could be observed: the British were conversant with torpedos anyway, and of course a good number ran aground and could be examined by the Allies. In addition, German propaganda made much of the existence of torpedos – and there was certainly no 'secret' of the acoustic detonation facility that many of them contained. There was nothing that the Allies could do – a totally silent ship was hardly practicable! – and the Germans liked to remind Allied forces of the might of the Third Reich and its power at sea, so great a challenge was it to the British reputation for naval supremacy.

The secret developments concerned the propulsion units (at first electric motors were used, but as we have seen there were startling new power plants available at the war's end) and also the homing devices which were being developed. One of the most successful projects was for a torpedo equipped with direction finding gear and servo-motor operated steering vanes which would home onto the sound of an enemy vessel's engines and explode on impact against the hull. This made the chances of success immeasurably greater and had this project been widely utilised by the Germans (particularly if the high-speed power units had also materialised in practice) then the Allied shipping lanes would have been closed altogether. British Intelligence, as it was, revealed the Germans' intentions to the Allies before the first of these experimental torpedos was ready for trial; 'squawker buoys' were ready in good time to confuse the homing apparatus. These were towed along by the Allied ship, making a loud noise which attracted the torpedo onto the wrong target.

Once again the Germans' plans were frustrated by events, and the loss of such a sophisticated device through the use by the Allies of a very simple and obvious gadget was a source of intense annoyance to the German

navy. Often it is the simplest of ideas which are the best, and this applies to both sides in conflict – in one case in particular it was such an elementary and inexpensive idea which gave the German U-boats a considerable advantage at sea. It was the *schnorkel*.

The design of this successful device, which enabled the U-boats to cruise under the water whilst drawing air enough to power their diesel engines – thereby recharging batteries without any substantial risk of radar detection or visual sighting of a conning tower – was another of Professor Walter's brainchilds. The first suggestion of the use of such a device, however, lies in history; for centuries it had occasionally been raised as a means of allowing men the freedom to negotiate the depths of the oceans. And indeed the first *schnorkels* were probably the top-secret Dutch devices which the Germans captured in the 1940 invasion of Holland.

However Professor Walter took the idea into a practical form, by designing and installing a *schnorkel*, fitted with a ball-valve to prevent sea water from lapping into it, in several submarines for experimental trials. The U-264, the first to have a *schnorkel* fitted, was later sunk by a British warship early in 1943, and a subsequent loss delayed the widespread use of the device. However by 1944 it was being fitted to more and more U-boats, and their range and usefulness increased accordingly, It was in August of that year, when the Americans began their drive towards Brittany, so cutting off the U-boats' safe passage to their anchorages at Brest, that the device really came into its own as the submarines were able to creep quietly in and out without being detected. Even so the *schnorkel* had disadvantages: when the ball valve was closed by a wave, the pressure in the hull – due to the continued running of the engines – fell suddenly, and oxygen from the atmosphere in the U-boat was replaced bit by bit by leaking carbon monoxide. Not only that but the *schnorkel* tube left a trail of bubbles, a significant wake, and it could be detected by airborne radar sets in searching allied aircraft.

Experiments carried out by Dr Meier-Windhorst at Draegerwerk showed that when the valve was permanently closed, the pressure in the submarine fell almost linearly to 600 millibars over a period of five minutes, as oxygen was consumed by the engines, at which time they all stalled. The drop was not considered to be serious to the crew during normal, short-lasting periods of closure caused by a high wave or a brief loss of attitude in the water.

By the end of the war this useful gadget had been fitted to a large number of German U-boats, and they proved to be valuable aid in the Germans' war at sea.

Communication was also a problem, and it was found that very long radio waves – about 28,000 metres – could penetrate through the upper layers of the ocean and, as long as the U-boat was within 50 feet of the surface, could exchange radio messages with its land base. The radio aerial responsible for the transmission of these unusual wavelengths was sited at Magdeburg.

And finally there was the problem of decoys: the Germans were keen to exploit any possible means of confusing the Allied forces in their search for the U-boats. Weighted buoys fitted with a tinfoil radar decoy were released from escaping U-boats, for example; by floating vertically in the water they created the same echo effect to the radar beam as the protruding conning tower of a submarine. By the time the searching aeroplane had reached the scene and found it deserted, the U-boat would be miles away, deep under the surface – and the buoy (which was constructed to sink after a short interval of time) had disappeared. This served to give the impression of a skilfully evasive U-boat and caused much confusion and, later, annoyance to the Allies.

Asdic (originally developed by a German scientist) was feared by the U-boat commanders above all: it meant the enemy was on the track and chances of escape were none too good.

Wire mesh grids were fitted to the sides of submarines at one stage, in order to provide interference sound-wave patterns calculated to confuse the tracking vessel. But substantially more successful were the bubble

Biber midget submarine:
Displacement: 6¼ tons. *Speed:*
6½ knots. *Range:* 130 miles. *Crew:* One.
Armament: Two 21-inch torpedoes

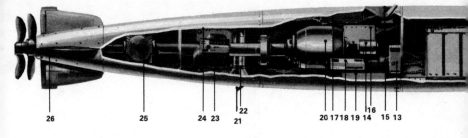

Type V German acoustic torpedo

1 **Acoustic receiver**
2 **Acoustic Amplifier**
3 **Thermal relay for safety range**
4 **Pick-up coils**
5 **Warhead**
6 **Solenoid locking pistol propellor**
7 **Contact (inertia) pistol**
8 **Coil operating pistol**
9 **Fusing relay**
10 **Pistol amplifier**
11 **Compressed air reserve**
12 **36-cell battery**

decoys carried by U-boats later in the war. These took the form of perforated cylinders of carbide which on contact with the sea water produced extensive, dense clouds of acetylene bubbles. These returned a vigorous Asdic echo and under the cover of this decoy the U-boat was able to make good its escape.

There is another application of the echo concept which has entirely different uses. The Germans, spreading ever wider the net of investigation and innovation, soon realised that the echo from the ground could be used by a flying aircraft as a means of measuring altitude. This gave rise to the top-secret sonic altimeter, a direct-reading device of unprecedented accuracy indicating the exact height – within feet – of a low-flying aircraft, far below the range where a conventional aneroidal altimeter could function accurately.

The device was produced at Atlas-Werke, Munich, and also, to a lesser degree, at the Luft-Hansel factory. Thousands were still on order at the end of the war, and only about 250 were ever fitted to aircraft. The device (named *Landehohenmesser*) operated by beaming a sonic 'ping' – like an Asdic transmitter – to the ground surface and then receiving the reflected echo from the surface. By timing the delay between the two events, the altimeter was able to indicate the height of the aircraft to within a foot or so of the true altitude. Because of the time delay in long transmissions, it was only useable below 500 feet, but it was found to be possible to fly aircraft on flat open territory at heights of almost exactly three feet!

The apparatus was in four main parts: the sound transmitter, the receiver, the electronic interpretation and display facility, and a source of compressed air (the last component being supplied by other manufacturers). In practice the air supply was from a petrol- or electrically-powered pump or from compressed-air bottles. The pump had to be capable of delivering 2 litres per second at a pressure of four atmospheres.

Alternatively, for smaller aircraft, an air bottle holding air at 120 atmospheres and with a volume of 8 litres was sufficient for two landings. The air was fed into a sound generator: quite simply, this was a whistle in the focus of a parabaloid reflector, which was tilted so as to throw the sound beam downwards and forwards, at an angle of ten degrees to the vertical, so that by the time it was reflected up from the ground surface it had moved forward to the region now occupied by the travelling aircraft. The whistle generated a note at 3,200 cycles per second – a high-pitched, shrill 'ping' – and the note was kept stable over a wide range of temperatures through

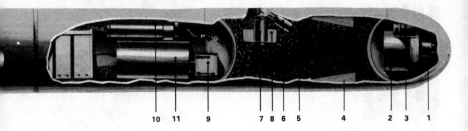

10 11   9   7 8 6 5   4   2 3   1

13 Main switch (motor circuit)	20 Motor
14 Charging plug	21 Touching lever switch
15 Starting lever	22 Depth control gear
16 Generator for homing gear supply	23 Gyroscope
17 G switch	24 Discriminator box
18 Converter for pistol supply	25 Contra-rotating gear
19 Pistol distributor box	26 Tail unit and propellors

the incorporation in the design of a bi-metallic strip which compensated for the alteration by changing the volume of the resonance chamber of the whistle itself.

The duration of the 'ping' was in the order of 0.015 seconds, the time being started and stopped by an automatic electro-magnet.

The receiver was in the form of a conical microphone diaphragm pointing ten degrees backwards and sensitive to the audible range of the transmitted beam of sound. Impulses were fed into an amplifier and timing unit which accurately discharged a small pulse of current proportional to the time taken for the sound to reach the ground and 'bounce' back up again; the current was fed into a moving coil milliameter reading 1 mA at full-scale deflection and calibrated in a scale of meters, from 0 to 150. Apparatus to suppress multiple echoes was included, in case a double-reflected sound beam was misinterpreted; and circuits to correct for passage of patches of temporary height (buildings in the landing path, for instance) and for background noise were also incorporated into the final design. There were disadvantages, such as the change of response when flying over a difficult surface (such as fresh snow) and the tendency for the slipstream noise of flying to cancel the usefulness out, at speeds above 190 mph; but

this was a most enterprising and useful device and would have assisted greatly in landing aircraft in difficult circumstances and in increasing materially the Germans' chances of striking back at the Allies.

Other workers were investigating recoil in weapons and finding ways of reducing it, so that larger calibre guns could be carried on aircraft and by men; plastics were being put to new uses and a consultant chemist, Professor Albert Schmidt of Konstanz, began to use fibreglass as a covering material for aircraft in the experimental stage. Thus the net spread wider: almost every facet of technology was examined, and a vast range of different approaches was brought to bear on the need for Germany to find new, secret weapons with which to win the war. But the Third Reich, in its muddleheaded introspection, left the holes in the net too large, and many of the most far-reaching of the developments that were caught up in it were allowed to slip through. Thus the fervour of nationalistic pride gave many men the impetus to discover new weapons, new processes and new techniques; whilst the blind manic blundering of the Nazi machine prevented their exploitation to the full.

Higher and faster- the secret jets, rockets and projectiles

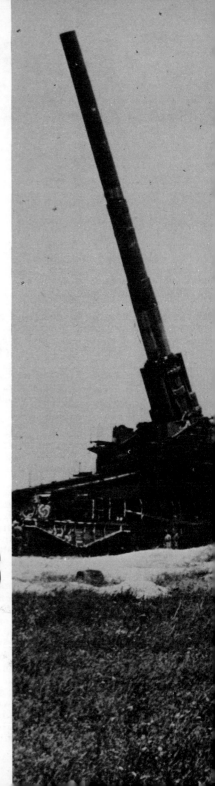

'Big Bertha' ready for action

The most obvious way of bombarding the enemy in wartime is by the use of guns, and Germany had many of these in variety. Some were small (such as the curious *Eckgewehr*, which had a curved barrel and fired around corners) and novel in design – but the range extended up to the biggest ever seen. At this the top end of the scale was the German 800-mm gun, named 'Dora' by the Germans and 'Big Bertha' by the Allies. This monster weighed over 1,000 tons and fired projectiles that were nearly a yard in diameter. The barrel was 90 feet in length and it was carried in two sections between firing sites, mounted on special railway wagons. For firing the entire monster weapon was assembled standing across two sets of rails. The crew for the entire operation totalled 1,500 men – the breech on its own weighed more than 100 tons – and it was capable of a maximum range of 54,000 yards.

But the range was not enough for operation against London, the most important target; and 'Big Bertha' could only be used against Sebastopol. It dawned on the operators that the best way of boosting the performance would be with rocket attachments, and two-stage rockets were designed to propel the charge after firing. Though this idea did not materialise in practice, it was a significant pointer to the lessons that the Germans were learning – lessons which centred on the use of rockets for long-range bombardment. Rockets are more complex and demanding than shells fired from guns, it is true; but their results are potentially more devastating by far and the Germans' progress in the field had already established them as world leaders in rocketry. So the eventual destiny of 'Big Bertha', the world's largest-ever cannon, was a symbolic pointer to the principle behind the Germans' most daring and sophisticated secret weapon – the guided missile.

In 1943 the *Enzian* was designed by Dr Konrad of Messerschmitts. It was in essence an unmanned version of the Me-163 *Komet* aircraft described earlier. This version was made of plastic wood and altogether twenty-five were produced and fired experimentally. Fifteen or so failed in flight, and the project was never put into full produc-

tion. The *Enzian* had four strap-on booster rockets which assisted the take-off.

Also inspired by another project was the *Schmetterling* rocket. This was adapted from the design of the Hs-293 glider bomb, and once again was fitted with four strap-on boosters for an assisted take-off. About sixty of this weapon (which some christened the 'V-3') were produced, although it never went into service; officially known as the Hs-117, it was produced at the Breslau factory of Henschel and was fired by two operators, one of whom controlled the flight of the airborne rocket with a miniature joystick connected to a radio transmitter which fed impulses directly to the servo-controlled steering fins of the missile. Thus the guidance operator could 'fly' the rocket as though he were sitting in the 'cockpit', just by moving the artificial joystick; simple as it seems the system was too filled with teething troubles to be very successful, and they were not ironed out by the time the war ended.

The *Rheintochter* was produced in two versions, Mark 1, 20 feet long, and Mark 2 only 17 feet; over eighty of the Mark 1 were fired, twenty or so with radio-controlled guidance – and the majority were very successful. In most cases radar was used to track the device, the impulses for direction control being fed directly from the radar station to the rocket. The Mark 3 was tested too, sometimes with strap-on rocket assisters for take-off. However the missile was only in the development stage when hostilities ended. Based on the experience gained in developing the *Rheintochter* 1 was the *Rheinbote*. This was a three or four stage missile with a relatively small payload (the warhead weighed only 88 lbs). Over 200 were used in battle, being fired against Antwerp in 1944 with considerable accuracy although, fortunately, with only limited damage being caused.

The *Feuerlilie* rocket was produced in two versions which were, in fact, quite different in many fundamental respects. The F-25 was 6 feet long and flew at subsonic speeds for 3 miles or so. It was developed at LFH Braunschweig, from bodies produced by the Ardeltwerke factory and motors from

An RAF officer examines a captured Henschel Hs-293 missile—the world's first air-to-ground guided weapon—with a 1,000-lb warhead

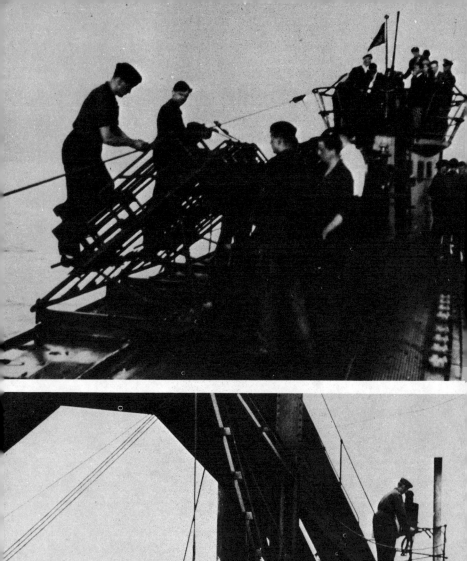

the Rheinmetall-Borsig in Berlin. It was first flown in Pomerania in 1943 but the programme was cancelled the following year.

At the same time the F-55 was begun. It was a more sophisticated missile by far: boosted with a solid-fuel first stage and thereafter powered by an alcohol/liquid oxygen mixture, it was over 15 feet tall and had a range double that of the earlier version. It was first tested in mid-1944, the second flight was not for six months (in December of that year) and the project was never realised afterwards. There was a little-known 'ghost' predecessor of the *Feurlilie* rockets, called the *Hecht;* it was launched on a sloping ramp and was 8 feet 2 inches long with a range of 6 miles and in some respects it was more like a miniature flying bomb, with a tailplane and swept back wings. But only a prototype ever flew before the *Feuerlilie* development pressed ahead.

Two interesting secret solid-fuel weapons were the BV-143 and BV-246, used against shipping. The intention was that the missile would fly down to near the sea surface and then, at an altitude of 10 feet or so, would straighten out and fly, wave-hopping, to the target. It didn't work at all well, and was soon abandoned. The 143 had a range of 10 miles and was 20 feet long; the 246 had a slightly longer range but was only 11 feet in length. Many homing systems were used, including advanced acoustic and infra-red sensitive devices and these experiments were more than useful to the Germans in later rocket developments – even though these particular rockets were eventually consigned to the scrap-heap.

A similar anti-ship weapon was the Fritz' – the SD-1400. It was produced in five or six different versions and was in essence an armour-piercing bomb fitted with wings. In terms of missiles employed per successful attack, the efficiency was small; but some of the ships sunk were useful gains to the German effort. The sinking of the battleship *Roma,* for

instance, became world famous in no time: this was carried out by a bomb launched from a Do-217 bomber.

A similar range of glide bombs commenced with the development of the Hs-293, launched from bombers and guided by radio to the target (a ship) by the navigator, who followed its progress visually and operated another 'joystick' control device. The bomb sank many Allied ships in the middle war period. It was followed on the production line by the Hs-294, 20 feet long (8 feet longer than its predecessor) and designed to end its journey as a torpedo, its wings breaking off as it entered the sea and thereafter homing acoustically to the motors of the enemy ship. The Hs-295 was a 16-foot rocket also designed as an anti-ship, bomber-launched missile which was cancelled shortly after going into production, and the Hs-296 was an experimental up-dated version of the series which was never fully tested in prototype stage. And finally there was the Hs-298, a light-alloy rocket also steered by radio impulses from a 'joystick' device operated in the parent bomber. It would have been a most successful anti-aircraft weapon, but was never fully developed. At 6 feet 8 inches long, and with a range of 5 miles, it weighed only 260 lbs and was an ingenious piece of design.

But there were many designs which emanated from the V-2, and which were never widely used in the war, or known about by the Allies until it was all over. These top-secret rockets included the *Wasserfall,* or C-2, in essence a scaled-down version of the V-2 in many respects, but with four stubby fins amidships. It was intended for use as a ground-to-air missile, but its test firings were far from successful: 25 per cent of the prototypes fired flew according to plan, the rest crashed shortly after launch. The final design would have boasted an infra-red homing device and a completely self-contained guidance system. The C-2 was 26 feet tall and had an effective range of 17 miles; its payload was a 675 lb bomb. From it in turn arose the tiny *Taifun,* just over 6 feet in length but capable of 2,800 mph – 1,000 mph faster than the C-2. It was fitted with either time-delay or proximity fuses and had no guidance system; it was

Left: Crew members prepare experimental surface bombardment rockets designed to be fired from a submerged U-boat

135

A rocket fired by a submerged U-boat during tests in 1942. *Right:*
The *Hochdruckpumpe* sectional gun showing the firing chambers

fired with the same purpose as anti-
aircraft barrage fire – simply to keep
the enemy aircraft at bay.

But the V-2 had also developed in
size on the drawing-board, and there
were inspired visions of monstrous
weapons of unprecedented potential.
The next in line had been the projected
A-6, a design for an improved fuelling
system making the V-2 idea more
stable in storage, instead of relying on
the hard-to-handle liquid oxygen as
oxidant. The A-7, instead of having
an improved fuelling system, had
wings attached for a final, prolonged
glide phase in its flight path; in this
way it was intended to increase the
range to double that of the V-2. The
A-8 was never constructed, either, but
the A-9/A-10 configuration was being
rushed into the prototype stage as
the war ended. It would have been a
vast two-stage rocket, nearly twice as
tall as the V-2 and with a huge new
booster that separated at 110 miles
from the winged version of an up-
rated A-4 which then accelerated away

and subsequently glided down to the
target, perhaps 3,000 miles away. It
was intended for the bombardment of
the USA.

But in the same way that – as in the
case of 'Big Bertha' – German gun
technology borrowed something from
rocketry, the exchange was also
reciprocated, giving rise to the use of
rocket-shaped shells fired from a gun
barrel. The idea was simple: fins on a
rocket stablise it in flight – so why
not fit a conventional shell with fins
and stabilise it in this way, rather
than by spin?

The result of this line of thinking
was the *Hochdruckpumpe* or HDP
(nicknamed variously *Fleissiges Lei-
schen* or *Tausendfussler*). This 'super-
gun' was an invention of the Saar
Rochling establishment and it took
the form of a very long sectional
barrel, with paired side-branches protruding at intervals like ribs on a
fish-bone. In these lateral firing chambers were explosive charges, fired
automatically in sequence. As the

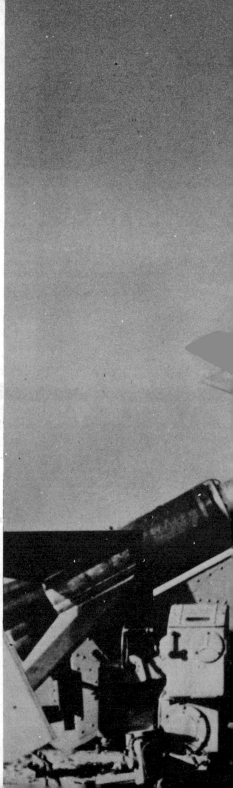

Right: the *Rheintochter*
anti-aircraft rocket on its launcher.
Above: Three shots from a sequence
showing *Rheintochter* being fired
during a test and streaking into the sky

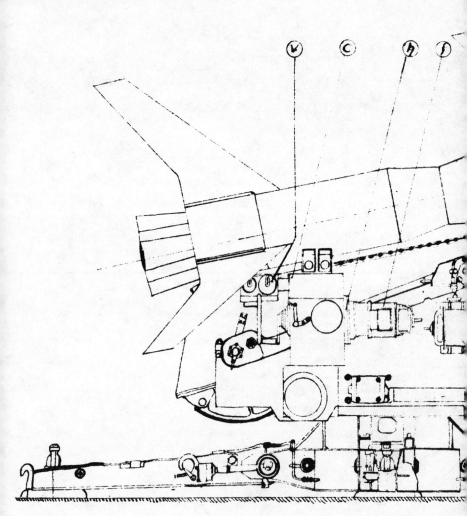

Rheintochter **R1**
**Two-stage flak rocket launched
from an inclined ramp.**
Altitude: **20,000 feet.**
Warhead: **250 lbs (with proximity fuse)**

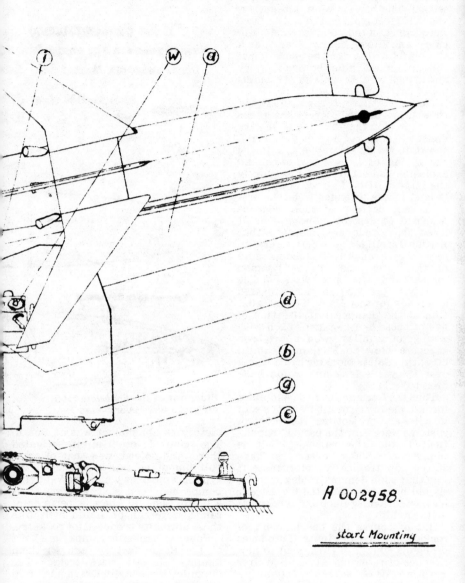

A 002958.

start Mounting

missile passed each of the branches the next charge in the sequence detonated, so building up the pressure until the missile was ejected at the formidable velocity of 4,500 feet per second. The barrels were, altogether, about 150 feet from end to end and two were constructed, one each at Antwerp and Luxembourg. There was a singular drawback to the operation of this unique machine, however; every few firings would result in the bursting of part of the barrel!

The operation of the 'super-gun' thus turned out to be a very hazardous procedure for the gun crew – but the sectional construction of the barrel, shown in the photographs, facilitated the fitting of a new part when an explosion had occurred. Subsequently the largest HDP of all was built near Calais; it was intended to be the first of a vast battery of such weapons designed to destroy London – or at least the Londoners' morale – but British Intelligence experts were informed by French resistance workers of the intentions of the German forces, and the site was bombed shortly before the gun was completed and ready for test firing. The installation of the planned initial ration of five of these 'super-guns' – each with a range of 85 miles – would have been a severe blow to England's capital city; this Calais monster had a barrel over 150 yards long and would have fired 150-mm shells.

Thus the Germans were developing a formidable array of different types of missiles and projectiles; their control systems were – for the period – sophisticated and the variety itself is impressive. Many observers have stated how tragically unfortunate it was that such demonstrable expertise should have been adopted by the war machine of a misguided and tyrannical dictatorship.

But of course the development of rockets and missiles in itself produces problems which, in turn, lead to new avenues of research. At the H Walter Kommanditgesellschaft, some interesting work on launching ramps (and other research of a similar nature) threw up a range of novel ideas. The commercial manager was *Korvetten Kapitan* Walter (the brother of Kiel's famous Professor Walter) and the

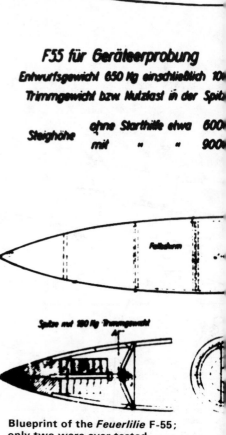

Blueprint of the *Feuerlilie* F-55; only two were ever tested

factory establishment was at Bosau. The central problem investigated there, and solved, was that of catapult launching ramps. Based on the type of ramp from which the V-1 flying bombs had been sent winging their way to London, these experimental ramps were run on liquid fuels like those normally produced for rocketry.

Four experimental ramps were built at the Bosau factory, and on them dummy aircraft were tested. The dummies were cylinders made from $\frac{1}{4}$-inch mild steel plate of low quality and roughly 10 feet long and 3 feet in diameter. Dummies ranging from 1,600 pounds to 7,700 pounds in weight had been tested and speeds ranging from 50 to 100 feet per second were

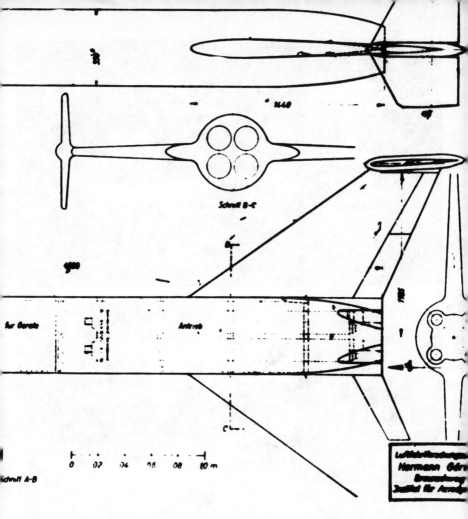

found after the body had travelled a distance of about 48-50 yards up the ramp. The speed measurements were all taken electronically. The ramps worked by having a piston forced along the firing cylinder at high speed, attached by an iron lever to the dummy itself; thus the steel cylinder (representing the aircraft) was ejected from the end of the ramp at the height of its acceleration. The device, stated the engineer in charge of testing (Max Mierke), was very promising. But, of course, the end of the war brought an end to this branch of research – at least for the Germans' purposes.

Rocket-powered torpedos were also studied here. Two years before the war's end an order for the production of the torpedos was put in hand; however they were not truly very successful. They burned a mixture of oil, hydrogen peroxide, a catalyst, and water (to keep the reaction temperature down); but the combustion was incomplete and the trail left by each torpedo was very marked indeed – also the maximum range was less than a mile. However when it came to starting the engine, it was often found that the ignition would take place not as combustion, but as complete detonation . . . and when this kept happening the naval authorities in Germany began rapidly to lose interest.

But this work in turn led to another avenue of exploration under the

Above: Scale models of *Schmetterling* (top) and *Wasserfall. Below:* Hecht, the predecessor of *Feuerlilie. Right:* Rheinbote

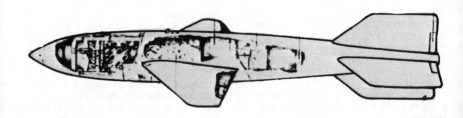

R. Spr. Gr. 4831
auf Meillerlafette

Rheinbote.

MASSTAB 1:100

205 1/44   5-Ausfertigung
2. Ausfertigung

KALIBER EILA: 191 cm
SCHUSSWEITE: 160 km
STREUUNG: NICHT BEKANNT
GESCHOSSGEWICHT: 1650 kg
davon 40 kg Sprengstoff
Vo = 1930 m/sec nach 10 km
TREIBSTOFF: 585 kg/Schuss
STARTBAHN AUF MEILLERWAGEN
LEISTUNG: 1 Schuss je Stunde
WIRKUNG: 15 - 21 cm Mörser
20 cm Granate
BAUFIRMA: RHEINMETALL BORSIG
ENTWICKLUNGSFACHMANN: Oberstleutnant AHA In.IV.
EINSATZ: ZUNÄCHST BLAFFETEN dazu 500 Schuss
AUFBAU FÜR ZIELWECHSEL: 15 Min

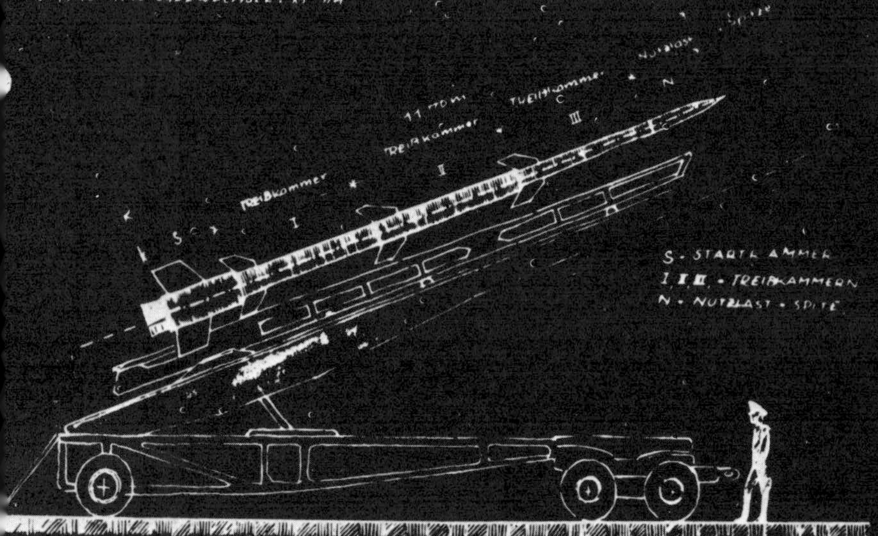

S = STARTKAMMER
I, II, III = TREIBKAMMERN
N = NUTZLAST + SPITZE

S. STARTKAMMER	I. TREIBKAMMER I	II. TREIBKAMMER II	III. TREIBKAMMER III	NUTZLAST
Leergewicht 470 kg	Leergewicht 240 kg	Leergewicht 220 kg	Leergewicht ... kg	Spitze
φ 535 mm	φ 267 mm	φ 267 mm	φ 191 mm	Leergewicht inkl.
240 kg Digl. RP 8e	240 kg Digl. RP 88	146 kg Digl. RP 88	46 kg Digl. RP 88	40 kg Teilen
fällt ab nach 3 km	fällt ab nach 12 km	fällt ab nach ... km	fliegt mit ins Ziel	
	RZ 1/10	RZ 5/10	RZ 5/10	Az

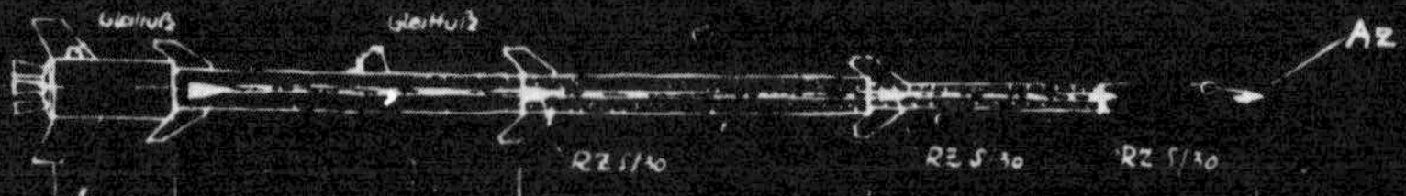

GESAMTGEWICHT: 1650 kg
GESAMTLÄNGE: 11,7 m

BERLIN, DEN 20. NOVEMBER 1944

SS GRUPPENFÜHRER UND
GENERAL. D. WAFFEN SS

leadership of Dr Oldenburg at the factory; a torpedo was invented which ran on a series of sharp, intermittent explosions: thus not only was it a successful reaction-driven torpedo, but it also set up shock waves of sufficient intensity to detonate mines in the area. The original plan was for merchant shipping to send out such a device (a farting torpedo, it was indelicately named by some) and then follow sedately in its protective wake.

But of course this use of rockets for the powering of torpedoes was rather freakish. Their prime application was, and indeed has been since, the propelling of rocket projectiles and high-speed aircraft. But there are serious drawbacks: the principle of the reaction motor (i.e. one which is driven forward by an engine producing thrust through burning gases) is obviously viable, but there are clear possibilities for an expansion of the concept. What if the oxidant was something less complicated than peroxide, for instance? The obvious oxidant is the oxygen in the air, and there are clear possibilities for using this, mixed with fuel, as a means of propulsion.

The answer, of course, was the jet engine.

The name associated with Germany's pioneer work is that of Professor Heinkel, who was concerned with turbine theory from the late 1930s onwards. It was in June 1939 that the He-176 was first test-flown, equipped with a Walter rocket motor, at Peenemunde; and shortly thereafter a Heinkel jet engine was fitted to a modified version of the aircraft to make the He-178. It flew for five or six minutes only near the end of August 1939, and was thus the world's first successful jet-propelled aircraft. Word was immediately rushed to the air ministry officials and thence to Hitler, all of whom were unimpressed. It was too much of a gamble when Germany was so near to winning the war, they felt, and the early success might just be a flash in the pan. So further research at government level was not considered, and Professor Heinkel had to go back to his drawing board alone and work on an improved model.

This turned out to be the He-162 and a steady programme of research began into its development, under conditions of total secrecy. However, the reticence of the German planners was such that the first aircraft to come off the production line did not enter service until 1945.

The He-162 had a wing span of 23.6 feet and a speed of over 500 mph with a range of up to 375 miles. But the Heinkel jet engines were not a great success in practice, and although they were fitted to both Heinkel and Messerschmitt aircraft at one stage, they were eventually superseded by engines produced by other companies.

So Professor Ernst Heinkel must go down in history as being the man to produce the very first successfully flown jet engine (it was code-named the He S/3)but he was not as successful as a development engineer as he was as a visionary mechanic.

One of the first jet engines to materialise as a successful enterprise in production terms was the Junkers 004 turbine. Its history dates from 1939, when at the Dessau Junkers factory a small model jet was designed. Its scaling-down resulted in experimental difficulties which came near to killing interest in the project at one stage: the turbines had a tendency to shed blades and in any event the reduction of the combustion end of the engine removed such a great proportion of the power that the compressor would not function adequately. But just before Christmas in that year, it was recognised that the failure might be due to the scale, rather than the operating principle, of the motor; and work was at once begun on a full-scale jet engine.

The engine was completed and ready for test runs before the end of the same year, a remarkable performance; it had eight compressor stages and six radial combustion chambers and was named the Ju-004/A. The tests showed that the engine could develop a thrust of over 1,750 lb and at the end of 1941 an engine was fitted to a converted Me-110 for aerial tests. Subsequently a number of modifications were carried out, mostly centering on slight improvements to the aerodynamic contours of the interior of the ducts and chambers and the improved engine the Ju-004/B – went into production by the end of 1943. That summer an Me-262 had been fitted for the first time with the 004/B engine and had shown considerable promise. But even so the production figures did not attain those set by the Luftwaffe: they agreed on a production rate of 1,000 per month in October 1944, but actually got 100; by December the agreed rate of production had gone up to 2,500 but it never exceeded 1,000 per month. However by the war's end over 5,000 of the engines had been supplied.

By using a neatly designed air-cooled, hollow turbine blade, thrusts of up to 2,200 lb were attained by the engine in use and the efficiencies were of the order of 80 per cent – good by any standards. After twenty-five hours of flying time the engines were stripped and examined; if they were found to be sound the engine was reassembled for a further ten hours of use, but then the engine was sent back for rebuilding as fatigue failure in the blades was a real risk by then. The engine tended to become unstable at altitude: but by carefully controlling the throttle an altitude of 42,000 feet proved eventually to be possible.

By the end of the war there were other projects at the Junkers plant, such as a design for a 17-foot-long engine with a thrust of 6,300 lbs running at 6,000 rpm which was intended to power a high-speed bomber. This was named the 012 engine – not because it was the second in a series, but because it was the custom to name the engines with a final code number which indicated the manufacturer. This code proved to be: 1—Heinkel Hirth, 2—Junkers aircraft, 8—BMW.

Thus the 012 was the first produced of its kind by Junkers, and there was no 011 – or, come to that, an 013 or 014. There was a plan for an 022, however; this would have been a modified 012 with a propeller fitted – a jet-prop engine, in other words. But design was never completed and the project did not even reach the prototype stage.

Later the 004 was boosted with afterburners but these were found to be unsuccessful: either they were wasteful of fuel or, if the fuel was

An *Enzian* anti-aircraft rocket ready for firing

148

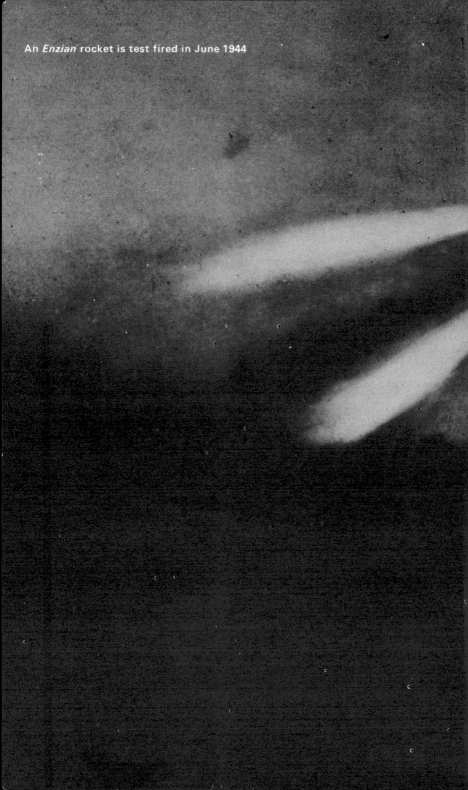

An *Enzian* rocket is test fired in June 1944

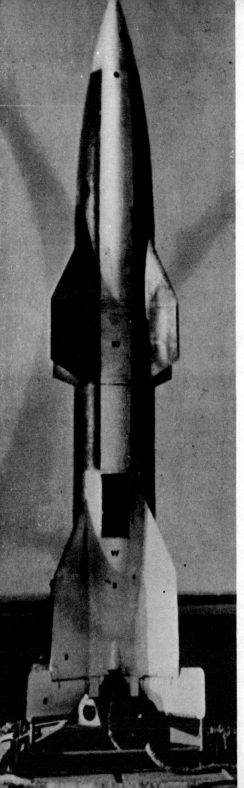

injected far enough forward to be effective, the turbine blades in the combustion chambers tended to distort and burn with consequent damage to the engine. Therefore this idea was soon abandoned.

The firm of BMW (Bavarian Motor-Works) was also involved with the development of Germany's top-secret jet turbines, at factories in Stassfurt, Heiligendroda, Mittelwork, and Eisenach under the chief control of Dr Scaaf, managing director, with a team of development experts named Stoffregen, Bruckman, Ostrich, and Knemayer. The research and development work was originally centred at Spandau, Berlin but with the bombing raids carried out by the Allies the production was moved out to other factories – some of them underground, in caves and salt mines.

There were 45,000 square yards of underground workshop, for example, in the salt mine at Eisenach and smaller units were situated at Abterroda, Plömintz, and Bad Salzungen – although this one, last on the list, never actually came into operation before the war's end.

At Nordhausen – where V-1s were being produced – another secret assembly line was established.

The development of the BMW 003 engine began with designs made in 1934; aerodynamic studies carried out at Göttingen suggested a design thrust of 1,320 pounds for initial trials and this version was built in 1939-40; on its first test run in August 1940 it produced a thrust of 990 pounds with an overall engine weight of 1,650 pounds, consuming air at the rate of 35 pounds per second. Later designs had a seven-stage axial compressor and hollow blades in the turbines were used here too – but mainly to economise in materials, apparently. Subsequent modifications put the thrust up to the region of 1,320 pounds.

It is interesting to note that the Luftwaffe, at this stage, showed no enthusiasm for the idea at all; when eventually they did opt for priority production of the Junkers engine the

Left: The *Wasserfall* heavy anti-aircraft rocket mounted on its firing stand
Right: Shortly after the launch

task of BMW was set as being the reduction of the consumption of raw materials: reputedly they eventually cut the nickel used in each 003 engine to only 1.32 pounds per engine by the end of the war. Meanwhile the labour cost of producing each engine was in the order of 600 man hours. Eventually the production schedules of the improved engine – rating up to 1,760 pounds thrust – were set at 2,000 per month at a whole range of production plants. But difficulties were encountered from the first, and eventually 1,500 per factory per month was decided upon — though even that had to be reduced after a few weeks.

The original application of the 003 was as an alternative for the 004 used in the Me-262 fighter described earlier, but production delays prevented that idea from materialising. Four of the engines were then envisaged as a power plant for an aircraft – the Arado 234 – with a speed of 550 mph at 32,800 feet and a range of 875 miles; indeed a prototype was constructed at Lünnewitz and was flight tested there before the war's end. The Arado firm brought out several small aircraft and there are reports that this latest of the line – the world's first turbine-propelled bomber – actually went into service in the spring of 1945. There were even more ambitious plans for a six-engined aircraft, the Ju-287, but this did not materialise.

An 018 engine was under development towards the end of the war: it was over 16 feet long and developed a thrust of 7,500 pounds at 5,000 rpm. It consumed air at the rate of 180 lb/sec and the turbine developed 40,000 hp (estimated). It was planned to instal this unit in a projected Henschel bomber – the P.122 – which would have had a design velocity of 635 mph and a range of 1,250 miles; but as the Allies advanced the development team placed gelignite charges in the fan blades of the turbine and blew the prototype (then under test) into pieces. Designs were found, and some damaged remains of the compressor were later taken to England for examination. It proved that the 028 would have been a most successful jet engine.

There were also similar, though less well-developed, plans under way at

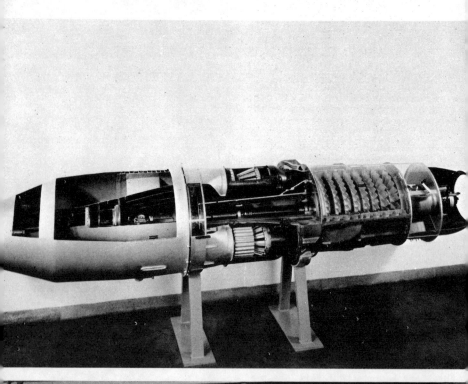

the Daimler Benz factory near Bad Harzburg where Professor Leist was in charge of research and development. He joined the firm at the outbreak of war after having studied jet theory for five years or so, and produced a prototype ZTL – meaning dual-circuit jet engine (*Zweikreis-Turbine-Luftstrahl*) – which had compressor and combustion system turbines mounted on drums which rotated in opposite directions, thus allowing higher temperatures to be reached and avoiding torque problems. There were two air flows, one through the main compressor and the second through a ducted fan which, fed into the turbines where combustion was taking place, served both to boost output and to cool the blades. It was felt that a gas temperature of 1,100 degrees centigrade could be reached by this method.

There are some interesting sidelights on the first jet plane to fly: as I said previously the first jet aircraft known to have flown took off for its maiden flight in August 1939. But were there, in fact, earlier experiments which still have not been recorded? I believe it is possible. In the first place, Dr Schaaf – Managing Director of the BMW organisation – was reported to have told the first Allied experts to occupy the plants during the liberation that 'the first jet aircraft to fly used a Heinkel-Hirth engine and flew in late 1937 or early 1938'. And Professor E M Evans, a distinguished chemical engineer who worked in Germany shortly before the war, described to me how a turbine motor – almost certainly a true jet – broke free from a test aircraft over Leipzig in 1938 and fell into the roadway 80 yards or so from where he was walking at the time. At that time there was no knowledge of what a jet engine was like – but rumours were heard occasionally, of secret research along these lines and in retrospect. Professor Evans is certain that the engine he saw was very similar to our now conventional concept of a jet engine. Several other people who were in Germany at that time (many of

them residents who still live in the same towns) recall having heard strange, whistling sounds coming from experimental aircraft. Many of these must have been test-bed flights in aeroplanes powered mainly by conventional motors. So perhaps the first true jet *did* fly in 1938, after all.

By 1944, much of the production work entailed in jet-engine production had been sub-contracted out to firms elsewhere, many of them in the Paris area. It was only as the Allies began to re-occupy France and the Low Countries that access to the information was obtained. Thus on 29th August 1944, an Allied team moved into Paris to begin to round up the available information. The Directeur des Industries Aeronautiques, M Blanchet, welcomed the liberation and at once threw open all available sources of information. But within a day or two the French attitude hardened once more: they were now less determined to give away all their own developments and more keen on preserving some national secrets. So the Allies were allowed only access to the purely German war-effort information, and techniques developed by the French were not disclosed.

But there really was not overmuch to release, when the matter was examined closely. As was the Germans' universal policy, they had told their French collaborators only enough to enable them to do the work, and no more; thus few of them had any idea of the source of their products or their eventual destination and, apart from gadgetry and fitments for the secret jet-engines, there was little to be found of any consequence.

But elsewhere in Paris there were some surprises in store for the advancing Allies: the Germans had been developing radar of a considerable degree of sophistication, and the range of devices under production in France was impressive. And – though radar does not quite fit into our brief in this book as a 'secret weapon' – there were several German radar training schools in France, too, which revealed how far they had gone in preparing for a new, electronic phase in the German war effort.

As we saw at the beginning of this book it was men like General Milch

Above: The Jumo 004/B jet engine.
Below: The world's first jet aircraft, the He-178. It first flew August 1939

Arado 234B
Twin-jet light bomber in limited use
by the end of the war.
Crew: One. *Span:* 46 feet 3½ inches.
Length: 41 feet 5½ inches.
Weight: 18,541 lb.
Top speed: 461 mph.
Armament: Two 20-mm cannon

He-162A-2
The 'people's fighter'.
Crew: One. *Span:* 23 feet 7½ inches.
Length: 29 feet 8 inches.
Weight: 5,840 lb.
Top speed: 522 mph.
Armament: Two 30-mm cannon

who laid most of the German war effort in the air. But the name which is associated with the effort more than any other is Albert Speer – it is surprising to reflect that he did not really emerge as a force in the *Reichsluftfahrtministerium* (literally: Reich's airtravel ministry) until mid-1944. He was a technical man more than anything, and soon put men like Milch in the shade by his brash efficiency and go-getting approach to war-time production. At once he set to reorganising and rationalising the effort and setting production quotas on an entirely different basis. Standard war-horses like the Ju-88 should be cancelled, he said, as they had out-worn their usefulness; new types of aircraft should be developed – and fast.

But Göring put his foot down and stopped many of Speer's ideas from advancing as fast as they would otherwise have done. Speer moved fast, but with torpidity from the Nazi overlords not fast enough to win the war – though too fast for Hitler. Milch, by this time interested only in keeping his job, was quick to agree with Speer wherever possible and when Hitler suggested that the Me-262 should be developed as a bomber, Milch was prepared to agree with Speer that the idea was nonsense.

The gist of Speer's argument was sound: the Me-262 was needed as a fighter, and had been designed as one; as a bomber it would be technically impressive, no doubt, but functionally inefficient. Milch agreed and – without informing Hitler that they were ignoring his wishes – the Me-262 was pressed ahead for production as a fighter. Hitler, in a megalomaniac rage, threw Milch out of office as soon as he heard the news; he forbade the very mention of the Me-262 as a fighter and curtailed its production. It was given bomb doors and a load of perhaps two bombs – which were completely useless as a tactical aid – and when, right at the end of 1944, Hitler realised his mistake and ordered (typically) that all the Me-262 bombers should at once be reconverted to fighters, it was too late. The Allies were advancing towards Berlin, France was completely lost, and the tide was running inexorably against Hitler and his foolish arbitrary decisions.

Meanwhile, realising the need for more and better fighters, the Führer issued a proclamation that a 'miracle-fighter' was to be developed. The plan was a panic measure. It took the form of a Jumo 004 engine mounted above the fuselage of a light wooden-and-metal fighter airframe (a configuration quite like that of the V-1 in many ways) and it was ordered by Hitler from the Heinkel works on 8th September 1944. There is a report that the first test flight took place on 10th December 1944, when a Captain Peter flew low across the airfield to impress the assembled Nazi and Luftwaffe representatives. The wing on the starboard side fell off suddenly, leaving the fighter cartwheeling crazily in the air for some seconds before crashing in pieces to the ground and killing the pilot. But there was no time for caution: the glue in the wings was thought to be at fault, the error was corrected, and at once production started and at the same time young men in the Hitler Youth began to enrol for training as pilots for the aircraft. After a week or two on gliders they were to be thrust aloft in a *Volksjäger* to fight the enemy – a terrifying fate for youngsters of perhaps seventeen or eighteen years old. But, fortunately for them, the war ended before this hasty project materialised to any extent. Less than a hundred of the aircraft were ever delivered, and most of those were grounded because of fuel shortages.

By early 1945 the fate of Germany's secret aircraft was sealed, and Göring began to lose all the remaining support he still enjoyed; indeed it was New Year's Day in that year which saw the last Luftwaffe attack of any size mounted against the Allies.

The V-rocket attacks were still to continue, of course, though the end of the war was by then clearly in sight. And so Hitler's vision of a range of secret, hard-hitting weapons evaporated into the mists of history and the German effort – so stalwart in some respects, but so muddled and confused in its direction as to be virtually impotent – literally fell over itself. It failed, like Hitler, the Reich itself, and the Nazi ideal.

But we would do well to remember how near it might have come to succeeding.

Bibliography

The Mare's Nest D Irving (Wm Kimber, London)
The Luftwaffe J Killen (Muller, London)
History of Rocketry and Space Travel W von Braun and F I Ordway
(Nelson, London)
German Secret Weapons of the Second World War R Lusar (Spearman, London)
German research in World War II, an analysis of conduct and research
L E Simon (Chapman & Hall, London and John Wiley Inc, NY)
The New Dark Age Brian J Ford (Purnell, London)
The Rocket Race Brian J Ford (Purnell, London)
Chemical Warfare Brian J Ford (Purnell, London)
Doctors at War Brian J Ford (Purnell, London)
Eyes in the Sky Brian J Ford (Purnell, London)
Germany's Secret Weapons Brian J Ford (Purnell, London)